Glaciation &
Periglaciation

Jane Knight

**Advanced
Topic***Master***

Series editor
Michael Raw

Philip Allan Updates, an imprint of Hodder Education, part of Hachette UK, Market Place, Deddington, Oxfordshire OX15 0SE

Orders

Bookpoint Ltd, 130 Milton Park, Abingdon, Oxfordshire, OX14 4SB
tel: 01235 827720
fax: 01235 400454
e-mail: uk.orders@bookpoint.co.uk
Lines are open 9.00 a.m.–5.00 p.m., Monday to Saturday, with a 24-hour message answering service. You can also order through the Philip Allan Updates website: www.philipallan.co.uk

ISBN 978-1-84489-617-2

First printed 2007
Impression number 5 4 3 2
Year 2012 2011 2010 2009

Printed in Spain

Hachette UK's policy is to use papers that are natural, renewable and recyclable products and made from wood grown in sustainable forests. The logging and manufacturing processes are expected to conform to the environmental regulations of the country of origin.

P01450

Contents

Impact

Introduction

The aim of this book is to provide a detailed and up-to-date review of glacial and periglacial environments for AS/A2 geography students. In keeping with other books in the Advanced TopicMaster series, *Glaciation & Periglaciation* provides a general understanding of processes and forms and is supported by detailed case studies. Activities interspersed throughout the text aim to deepen knowledge and understanding and develop essential investigative skills, including the statistical analysis of data.

Glaciers are among the most powerful agents shaping the surface of continents. Through erosion and deposition they create imposing landscapes that have long fascinated geomorphologists. However, glaciation has more than just academic interest: glaciers and their landscapes are important resources, supporting tourism and providing hydroelectric power and water supplies for millions of people. Periglacial landscapes may be less dramatic than their glaciated counterparts, but are none the less interesting. They are mainly controlled by the growth and decay of small bodies of ice. Vast areas of the northern hemisphere experience periglacial conditions. Their potential for economic development is considerable but raises important issues.

Today, both glacial and periglacial environments are under enormous pressure. Conflict between developers and indigenous tribal groups occurs in the Arctic tundra; there is pressure on fragile glacial and periglacial environments from unsustainable tourism and energy production. Meanwhile, nowhere is climate change occurring more rapidly than in glacial and periglacial environments. The implications of these changes are global and threaten the lives of hundreds of millions of people during the twenty-first century.

Jane Knight

1 An overview of glaciation

Why study glaciers?

Glaciers are among the most powerful natural forces acting on the Earth's surface (Figure 1.1). Their importance extends beyond their sheer size as they are also:

- significant agents of erosion, which have created some of the most spectacular landforms on this planet
- our most important store of fresh water (75% of the world's fresh water is locked up in ice sheets)
- able to modify global environmental systems:
  - the circulation of the oceans has been altered previously by the growth and, more importantly, the decay of glaciers
  - some ice sheets become so extensive, they modify the global climate and lower temperatures
- an important control on past and future global and regional changes in sea level
- responsible for redistributing minerals and other natural resources

Figure 1.1 **Examples of glaciers**

(a) Franz-Josef glacier, South Island, New Zealand

(b) Solheim outlet glacier of the Myrdalsjokull ice cap, Iceland

(c) Gulkana glacier, Alaska

R.March, US Geological Survey

(d) Holgate glacier, Alaska

B. Molnia (2004) *Holgate Glacier*: from the *Online glacier photograph database*, Boulder, Colorado USA: National Snow and Ice Data Center/World Data Center for Glaciology: digital media

(e) South Cascade glacier, Washington state, USA

B. Krimmel, US Geological Survey

(f) Aletsch glacier, Switzerland

D. Beyer/Pictures

Activity 1

Use the internet to research the current importance of glaciers and how they are of significance for the future. The BBC news website pages listed here are a good starting point for this exercise:

news.bbc.co.uk/1/hi/sci/tech/4471135.stm

news.bbc.co.uk/1/hi/uk/4228411.stm

news.bbc.co.uk/1/hi/sci/tech/3662975.stm

Extent of glaciers and ice sheets

Glaciers are dynamic systems. The coverage of the Earth's surface by ice is known to have changed significantly during the **Quaternary** era (the past 2.3 million years) in response to climatic warming and cooling. This has resulted in a series of **glacials** that occurred during the **Pleistocene** epoch, with extended ice coverage and **interglacials,** where ice coverage was often less than it is today (Figure 1.2).

These two phases have been punctuated with either **stadials** (short-lived cold climatic episodes that occur during a predominantly warm phase) or **interstadials** (short-lived warm phases that occur during a relatively cool climatic phase).

| Figure 1.2 | The ocean signature of climate change — the oxygen isotope record of ocean sediments clearly shows periods of intense cold and warmer interglacials (data collected from the Greenland ice sheet) |

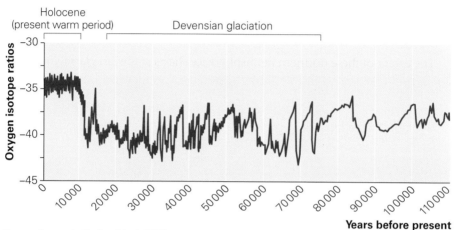

Source: *Geography Review* (March 2000)

The most important climatic variations in the British Isles during the Quaternary era are summarised in Table 1.1. However, there remains some uncertainty about the precise number of glaciations experienced in the British Isles over the past 2 million years.

| Table 1.1 | Chronology of glaciations and interglacial events during the mid-to-late Pleistocene |

Time	Epoch/stage	Climatic characteristics
Present day to 10 000 years BP	Holocene	Current interglacial; average global temperatures 4°C above those in last glaciation; global warming a potential threat
10 000–26 000 years BP	Pleistocene/late Devensian glaciation	Glaciation; maximum extent of ice sheets reached at 18 000 years BP
26 000–115 000 years BP	Late Pleistocene/early and mid-Devensian glaciation	Onset of Devensian glaciation
115 000–128 000 years BP	Ipswichian	Interglacial
128 000–175 000 years BP	Mid-Pleistocene	Glaciation; the occurrence of this is disputed
175 000–375 000 years BP	Mid-Pleistocene	Hoxnian interglacial
375 000–500 000 years BP	Mid-Pleistocene	Anglian glaciation; ice reached the Thames Valley

Just 18 000 years ago, the last glacial maximum was reached. This was when the ice was at its greatest extent and much of the northern hemisphere was buried under extensive ice sheets:

- The Laurentide ice sheet extended from the Arctic, with its southern margin in what is now New York State. The ice covered 75% of Canada (the other 25% being covered by the Cordilleran ice sheet). It reached a thickness of 4 km.
- Scandinavia was covered by the Scandinavian ice sheet, which extended across the North Sea to the British Isles and reached as far south as London.
- The Greenland ice sheet was significantly larger than it is today.

The extent of these northern hemisphere ice sheets at the last glacial maximum is shown in Figure 1.3, while Table 1.2 summarises their volume and extent.

Figure 1.3 **Extent of the northern hemisphere ice sheets at 18 000 years BP, the last glacial maximum**

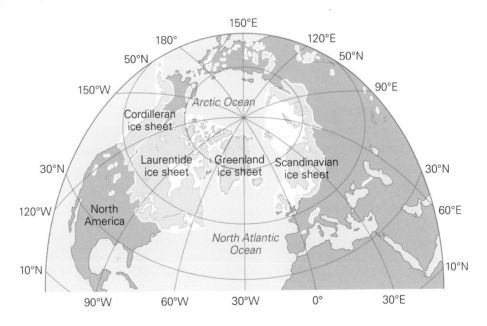

Table 1.2 Dimensions of former ice sheets and present-day Antarctica

Ice sheet	Area	Present volume	Devensian maximum
Laurentide ice sheet (North America)	10.2–11.3 × 10⁶ km²	0	34.8 × 10⁶ km³
Greenland ice sheet	1.7 × 10⁶ km² (present-day size)	2.4 × 10⁶ km³	3.5 × 10⁶ km³
Antarctica	14 × 10⁶ km² (present-day size)	30 × 10⁶ km³	34 × 10⁶ km³

Why do glaciers develop?

The ice in the last glacial (the Devensian) extended to relatively low latitudes — areas that today are too warm for ice to accumulate. So what caused the development of such enormous bodies of ice? And, of even greater interest, how did glaciers survive at such low latitudes? Several theories have been suggested to explain the growth (and decay) of glaciers:

- From observations of striations and landforms on the glacier in Zermatt, Switzerland, Louis Agassiz (1840) was among the first scientists to suggest that glaciers had previously been far more extensive than they are today.
- Lyell (1841), who was initially an opponent of Agassiz (but later became a convert), lectured about the 'Geological evidence of the former existence of glaciers in Forfarshire' to the Geological Society of London.

By the mid-nineteenth century, thanks to the observations of Agassiz, Lyell and other geologists, the scientific establishment had started to accept that there had been episodes in the Earth's history when glaciers had been far more extensive. However, there was no mechanism given to explain the growth and decay of glaciers on a continental scale. Milutin Milankovitch, a Serbian mathematician, changed all this.

Milankovitch calculated that the astronomical relationship between the Earth and the sun varied over time and so gave rise to warmer and cooler climatic episodes. He assumed that the amount of energy emitted by the sun remained constant and argued that another factor was responsible for the cooling that caused ice ages to develop. The key was changes in the Earth's orbit and tilt over three different timescales:

- the Earth's orbital path around the sun (eccentricity)
- changes in inclination/obliquity
- the precession of the equinoxes

Eccentricity

Milankovitch's calculations showed that over a period of about 100 000 years, the Earth's orbital path around the sun (eccentricity) changed from elliptical to circular and then back to elliptical. This cycle exaggerated the differences between seasons and altered the amount of solar radiation received by the Earth by as much as 30%. Currently the Earth is closer to the sun during winter in the northern hemisphere, hence our winters are mild, and is at a greater distance during summer, so we are experiencing relatively cool summers. Figure 1.4 shows the changing orbital path of the Earth around the sun, from circular to elliptical.

Figure 1.4 **The changing path of the Earth's orbit around the sun (eccentricity)**

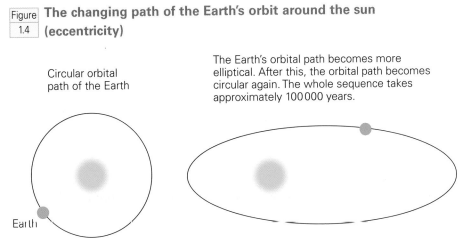

Circular orbital path of the Earth

The Earth's orbital path becomes more elliptical. After this, the orbital path becomes circular again. The whole sequence takes approximately 100 000 years.

Earth

Changes in the angle of the Earth's axis

The Earth is currently tilted at 23.5° but, as Milankovitch calculated, this varies every 41 000 years between 22.1° and 24.5°. The greater the tilt, or obliquity, the longer the winter in the northern hemisphere, which results in lower temperatures and means that ice is able to accumulate (Figure 1.5).

Figure 1.5 **Changes in the Earth's tilt (obliquity). This cycle occurs every 41 000 years.**

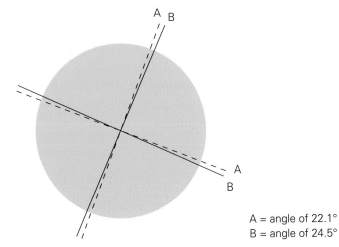

A B

A
B

A = angle of 22.1°
B = angle of 24.5°

Precession of equinoxes

The precession of the equinoxes has a periodicity of 21 700 years. Milankovitch calculated that the Earth's axis of rotation 'wobbles' like a child's spinning top as the Earth rotates round the sun. This means that the time of year at which the Earth is closest to the sun varies. Milankovitch calculated that the Younger Dryas (mini ice age) of 11500 years ago occurred because the northern hemisphere was pointing away from the sun and so temperatures were lowered. Figure 1.6 shows the wobble of the Earth's axis of rotation.

| Figure 1.6 | **Every 21700 years, the Earth's axis of rotation 'wobbles'** |

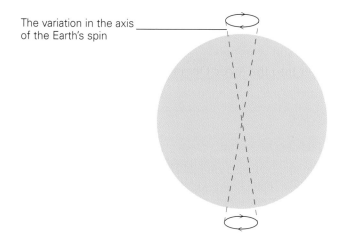

The variation in the axis of the Earth's spin

Milankovitch was able to use the combination of these three models to show how the receipt of **insolation** over the past 600000 years has varied. Of even greater importance was the correlation of insolation values with the timing of known glacial periods.

Influence of the oceans

The importance of oceanic circulation has been revealed recently. Scientists have demonstrated how the North Atlantic Drift (NAD) partly controls temperatures in western Europe. The NAD is a unique current of water in that it transfers more warm water to higher latitudes than any other ocean current.

The NAD can be switched on or off. Currently it is switched on and it transfers warm waters from the Gulf of Mexico northwards across the Atlantic Ocean to northwest Europe. The water is warm and relatively low in salinity and therefore of low density. As it moves northwards, it cools and increases in salinity and density. Eventually the denser water sinks and returns at depth towards the equator (Figure 1.7). However, any reduction in the salinity will switch off the conveyor and prevent the transfer of warm waters from the Gulf of Mexico to western Europe.

How can the NAD conveyor be switched off? Basically, through a large influx of fresh water, caused in several ways:

- The melting of northern ice sheets, as seen at the end of the last glaciation. This increases the volume of fresh water injected into the North Atlantic.
- The production and melting of icebergs, as seen at the end of the last glaciation from the Laurentide and European ice sheets. These are known as Heinrich events, when icebergs drifted across the North Atlantic from the Hudson Strait and melted. Scientists think that the melting of icebergs contributed to the shutdown of the NAD, which led to cooling and then the Younger Dryas stadial of 11 500 years ago.
- Large inputs of fresh water into the Arctic Ocean from Siberian rivers.

Figure 1.7 **The path of the North Atlantic Drift oceanic current**

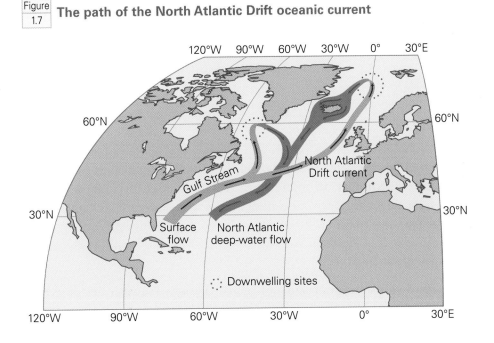

Classification of glaciers

There are several different types of glacier. They can be divided on the basis of their size or whether they are land-based or marine-based. Glaciers can be **constrained** by valley sides, or **unconstrained**, in which case they flow freely over the surrounding land. Regardless of their type, all glaciers flow and can transform the landscape by erosional or depositional processes.

Unconstrained glaciers

Unconstrained glaciers tend to be the largest glaciers. They exist where ice is so thick and extensive it submerges the landscape. They are known as **ice sheets** and are drained by **ice streams** and **outlet glaciers**. For example, present-day **Antarctica**, the largest ice sheet on Earth, is approximately 58 times the size of the UK. It has an average thickness of 1829 m and a maximum thickness of 4776 m.

A land-based ice sheet drained by ice streams covers 88.74% of Antarctica. The East Antarctic ice sheet is land based but the West Antarctic ice sheet rests predominantly on the sea bed, which is at a depth of 2500 m in places. Other contemporary examples of unconstrained glaciers are the **Greenland ice sheet** and the Icelandic ice field, the **Vatnajökull**, the largest glacier in Europe (8100 km²), which is drained by large outlet glaciers, such as Breidamerkurjökull and Skeidararjökull.

Constrained glaciers

The extent of constrained glaciers is defined by topography. For example, **valley glaciers** are constrained laterally by the valley sides. Other constrained glaciers are **cirque glaciers**, **piedmont lobes**, **niche glaciers** and **outlet glaciers**. Examples of constrained glaciers, and in particular valley glaciers, are widespread. Well-known examples are found in the Alps (e.g. Mer de Glace) and Alaska (e.g. Gulkana glacier and Wolverine glacier) and include South Cascade glacier in Washington state. Constrained outlet glaciers, such as Breidamerkurjökull and Skeidararjökull, drain from the Vatnajökull ice field in Iceland — the ice field being unconstrained.

Activity 2

Use the photograph archives of the National Snow and Ice Center, USA, found at www.nsidc.org/data/glacier_photo/photo_query.html, to find examples of the different types of constrained glacier.

Land-based and marine-based glaciers

Land-based glaciers are those in which the base is at or above sea level. Glaciers in Iceland, the Rockies, the Alps and the East Antarctic ice sheet, for example, are land-based. So too were the Laurentide and Scandinavian ice sheets.

A marine-based glacier is one in which the base is below sea level. For example, the West Antarctic ice sheet is marine-based, with its base being up to 2000 m below sea level (frozen to the sea bed).

The glacial system

Glaciers are dynamic systems with inputs, processes and transfers, and outputs (Figure 1.8).

| Figure 1.8 | **Flow diagram summarising the glacial system** |

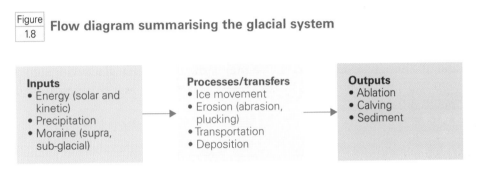

The balance between the inputs and outputs of snow and ice in a glacier is called the **mass balance**. This can be either **positive** (when there are more inputs than outputs and so the glacier **snout** advances) or **negative** (when the outputs exceed the inputs and so the glacier snout retreats). It is important to remember that while the snout advances and retreats, the ice is still moving under its own weight and gravity.

Zones of a glacier

A glacier can be divided into two zones where:
- **accumulation** exceeds ablation
- **ablation** exceeds accumulation

The former is known as the **zone of accumulation**, where precipitation is converted into glacier ice. The latter is called the **zone of ablation**, where melting predominates. The two zones are separated by the **equilibrium line altitude (ELA)**, which is where there is a balance of inputs and outputs. This usually occurs about two-thirds of the way down a glacier (Figure 1.9). Its location depends on the balance of inputs and outputs. The volumes of accumulation and ablation are quantified in **metres of water equivalent** to standardise measurements.

Figure
1.9
Zonation of a glacier

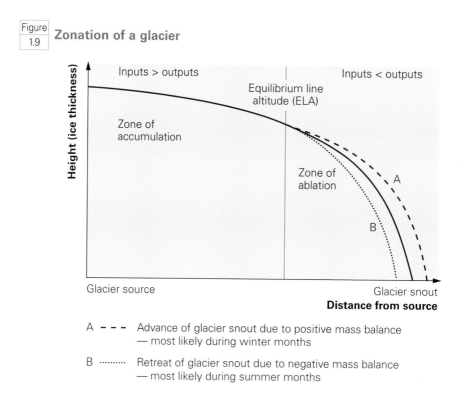

A − − − Advance of glacier snout due to positive mass balance
— most likely during winter months

B ·········· Retreat of glacier snout due to negative mass balance
— most likely during summer months

The **balance year** follows a pattern of increased glacial volume (in metres of water equivalent) during the winter months, and decreased glacial volume during the summer months. This is summarised in Figure 1.10.

Figure
1.10
The balance year for glaciers

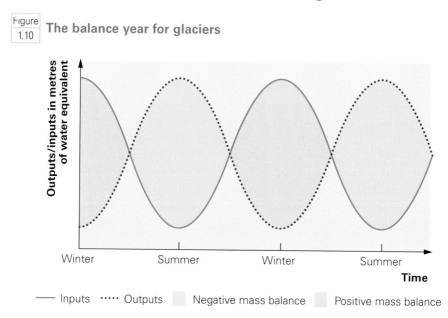

Activity 3

Table 1.3 shows the mass balance for the Gulkana glacier in eastern Alaska.

Table 1.3 Winter, summer and net mass balance for the Gulkana glacier in eastern Alaska, 1974–2004

Year	Winter (metres of water equivalent per year)	Summer (metres of water equivalent per year)	Net balance (metres of water equivalent per year)
1974	0.55	−1.64	−1.09
1975	1.12	−1.35	−0.23
1976	0.96	−1.88	−0.92
1977	1.39	−1.60	
1978	0.98	−1.17	
1979	1.35	−1.87	
1980	1.13	−1.20	
1981	0.96	−0.93	
1982	1.55	−1.67	
1983	1.14	−1.11	
1984	1.30	−1.61	
1985	1.41	−0.73	
1986	1.09	−1.03	
1987	1.24	−1.37	
1988	1.26	−1.48	
1989	ND	ND	–
1990	1.36	−2.04	
1991	1.31	−1.37	
1992	0.98	−1.22	
1993	0.82	−2.49	
1994	1.37	−1.96	
1995	0.94	−1.65	
1996	0.87	−1.39	
1997	0.99	−2.68	
1998	0.79	−1.43	
1999	1.04	−2.15	
2000	1.44	−1.49	
2001	1.40	−2.08	
2002	0.76	−1.83	
2003	1.79	−1.80	
2004	0.93	−3.22	

Source: US Geological Survey

Development

Activity 3 (continued)

(a) Calculate the net balance for each year (the first three are done for you).

(b) Use an appropriate technique to present the data showing the winter and summer mass balance.

(c) Describe the changes shown by the glacier.

Accumulation

Accumulation is due to the precipitation of snow and the formation of glacial ice. Rates of glacial ice formation depend on atmospheric temperature and the amount of precipitation. There are several stages in the formation of glacial ice. When snow falls onto a glacier, it traps atmospheric gases in large pore spaces between the snowflakes. Snow has a low density of $50-70 \, kg \, m^{-3}$. The tips of the snowflakes melt and the snow begins to compact. The density increases to $400-830 \, kg \, m^{-3}$. The snow becomes more granular as the flakes become more rounded.

This transformation of snowflakes is relatively quick and typically occurs over a few days in temperate areas such as Greenland. In Antarctica, the transformation is much slower, taking several years. Where this granular snow exists over a summer with little or no melting, it is known as **firn** or **neve**. The bottom of the firn is demarcated by the impermeable ice below.

The final stage in the transformation of firn to glacier ice is greater compaction during which the bubbles of air trapped in the ice are slowly reduced in size. The density can increase to $910 \, kg \, m^{-3}$. Depending on the environment, the timescale for transformation from firn to ice will vary:

- In Switzerland, firn can exist for over 12 years with glacial ice often only forming after 25–40 years.
- Greenland requires 150–200 years for the transformation or firn into ice to occur.

The timescale of the transition from snow to ice varies. For example, glacier ice at Vostok in Antarctica takes about 4000 years to form, whereas in the milder and more humid Arctic it takes only 100 years.

Another considerable source of input material is the snow, ice and rock derived from avalanches onto the glacier surface.

Ablation

Ablation is the loss of mass from the glacier. While most evidence of ablation can be seen as **meltwater** at the snout of a glacier, some ablation also occurs due to evaporation from the surface of the glacier. Evaporation accounts for about 0.5% of ablation.

The surface **albedo** (reflectivity) is an important influence on ablation. Snow has a high albedo, with 60–90% of short-wave radiation being reflected from the surface. Glacier ice has a lower albedo, with 20–40% of short-wave radiation being reflected. Thus, there will be greater ablation on a less reflective ice surface compared with more reflective snow. **Supraglacial debris** (rocks that rest on the surface of the glacier) can influence this locally as it provides some insulation, causing the surface to melt unevenly.

Another important way in which some glaciers and ice sheets lose mass is by **calving**. This occurs where the snout ice margin extends into water — either the sea or a lake. Calving is the loss of ice at the snout through the production of icebergs. Two of the greatest margins vulnerable to calving are the Filchner-Ronne ice shelf and the Ross ice shelf of the West Antarctic ice sheet. These are both floating margins. Between January and March 2002, Larsen ice shelf B discharged thousands of icebergs into the Weddell Sea. The total area of the ice discharged was 3250 km², which is an area larger than Luxembourg.

2 Why do glaciers move?

The rate, amount and type of movement of a glacier are determined by several factors, most notably the **thermal regime** of the glacier and the **gradient** of the ice surface. Over time, the type and speed of flow of a glacier changes. This is due partly to the balance between inputs and outputs. Consequently, a glacier may experience phases of relatively rapid flow known as a **glacial surge**. Some sections of an ice sheet might flow very rapidly. This is known as an **ice stream event**.

Thermal regime of glaciers

Warm glaciers

Warm glaciers are found in Alpine and sub-Arctic areas, although it is now known that some parts of the Arctic and Antarctic ice sheets are warm based. Ice is at or close to **pressure melting point (PMP)** throughout the warm glacier (Figure 2.1(a)). This means that, although temperatures are below freezing, water can exist. This is due to:
- the weight of the overlying ice creating high pressure, which warms the ice
- friction with the valley sides and bed which is great enough to melt the ice

These processes generate meltwater during the warmer summer months, although temperatures are cold enough in the winter for no melting to occur. The vertical profile of temperature within a glacier or ice sheet decreases to a certain depth and then, with the influence of pressure melting or **geothermal heat flux** (from the Earth's interior), it slowly increases.

A rather surprising example of a warm-based thermal regime can be found in parts of Antarctica. Here, geothermal heat flow helps with basal melting of ice. Up to a third of Antarctica is now considered to be warm based and at pressure melting point. Until recently it had always been assumed that Antarctica was entirely cold based. However, improved technologies have shown the presence of subglacial meltwater in the heart of Antarctica. Thus it is possible for water to exist and flow at subzero temperatures.

Cold glaciers

Cold-based glaciers are frozen to their bed (Figure 2.1(b)). Two-thirds of Antarctica's glaciers are considered to be cold based. There are two conditions that lead to the formation of cold-based glaciers:

- The air temperature when firn and glacier ice were formed was below freezing.
- Air temperatures throughout the year are so low that there is no melting of the surface layers.

Figure 2.1 **Change of temperature with depth in (a) warm- and (b) cold-based glaciers**

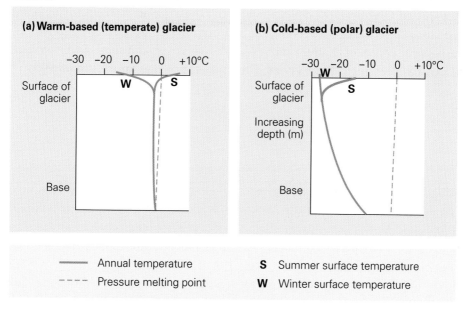

(a) Warm-based (temperate) glacier

(b) Cold-based (polar) glacier

—— Annual temperature	**S** Summer surface temperature
---- Pressure melting point	**W** Winter surface temperature

Activity 1

Research examples of warm-based and cold-based glaciers using textbooks and the internet. Complete a copy of Table 2.1.

Table 2.1 Examples of warm-based and cold-based glaciers

Name of glacier	Thermal regime	Location

Movement of ice

The **basal thermal regime** of glaciers is a key characteristic that partly determines the type of movement and speed at which glaciers flow. At different depths, ice is subjected to different amounts of pressure:

- Ice behaves like a plastic and deforms in order to move at greater depths.
- The surface is subject to less pressure; the ice is more brittle and so tends to fracture due to movement.

For all types of movement, one of the most important forces on the ice is the stress that is applied. This can be **normal stress** or **shear stress**. Normal stress is the force exerted on material perpendicular to the surface. Shear stress is the force that moves the ice down hill and is parallel to the surface (Figure 2.2).

Figure 2.2 **Normal and shear stress. Stress is important for determining movement of a glacier.**

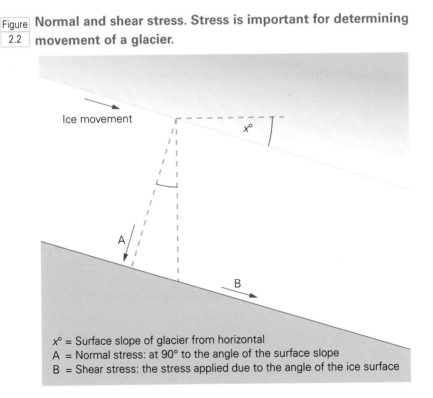

$x°$ = Surface slope of glacier from horizontal
A = Normal stress: at 90° to the angle of the surface slope
B = Shear stress: the stress applied due to the angle of the ice surface

Glaciers tend to flow parallel to their surface gradient. Movement is driven by gravity and there is resistance to this from friction with the bed and valley sides. Thus the slowest moving parts of the glacier are usually the valley sides and valley bed, while the fastest flowing areas are towards the surface and in the centre (Figure 2.3).

Figure
2.3 **Velocity profile (a) across a glacier and (b) with depth**

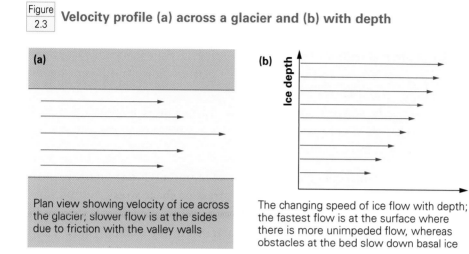

Plan view showing velocity of ice across the glacier; slower flow is at the sides due to friction with the valley walls

The changing speed of ice flow with depth; the fastest flow is at the surface where there is more unimpeded flow, whereas obstacles at the bed slow down basal ice

Mechanisms of ice movement

There are several different mechanisms of ice movement:

- internal deformation (ice creep)
- basal sliding
- subglacial deformation

Internal deformation

Internal deformation (**ice creep**) is a very slow movement of ice, which is due to the **polycrystalline** nature of ice (i.e. composed of many crystals). Movement is by the dislocation of individual crystals and not of a whole body of ice, hence its slow speed. The bottom of the ice crystal remains stationary and this forms a base for movement. Gravity moves the top of the ice crystal downhill (Figure 2.4).

Figure
2.4 **The process of ice creep on an ice crystal**

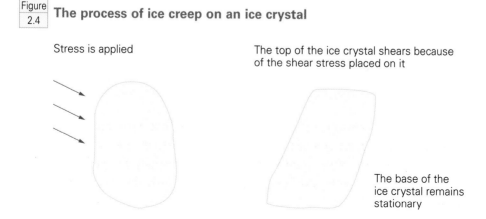

Stress is applied

The top of the ice crystal shears because of the shear stress placed on it

The base of the ice crystal remains stationary

Glen's Flow Law determines the amount of movement that occurs by ice creep. Glen's Flow Law states that if shear stress is doubled (for example, by an increased gradient), the creep will increase by a factor of eight. This explains why so much movement occurs in the **sole** of cold-based glaciers.

To calculate Glen's Flow Law, the shear stress (τ) placed on the ice must be calculated first:

$$\tau = \rho g h \sin\alpha$$

where

τ = stress

ρ = ice density

g = gravitational force

h = the thickness of ice above the point stress is being calculated for

α = the gradient of the ice surface

The stress is then a component of Glen's Flow Law and the following is used to calculate glacier movement (strain):

$$E = A\tau^3$$

where

E = strain

A = a figure that reflects the hardness of the ice

τ = stress

This means that the amount of stress applied to different types of ice will determine the amount of internal deformation that occurs (Figure 2.5).

| Figure 2.5 | **Glen's Flow Law** |

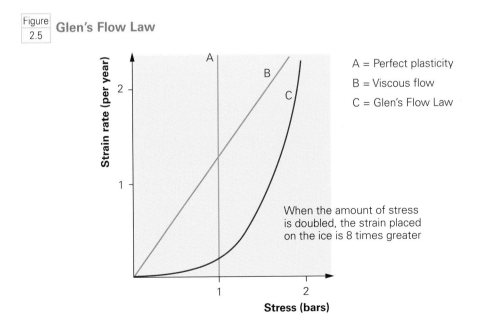

A = Perfect plasticity

B = Viscous flow

C = Glen's Flow Law

When the amount of stress is doubled, the strain placed on the ice is 8 times greater

Basal sliding

Basal sliding occurs where the base of a glacier slides over the surrounding sediments and rocks. The two processes associated with this are **basal creep** and **regelation slip** (Figure 2.6).

Basal creep

Obstacles on the bed of a glacier promote basal creep. Pressure increases on the up-ice side of an obstacle causing ice crystals to deform around it. You should note that the larger the obstacle, the greater the pressure needed to negotiate it, and so the greater the amount of deformation. The ice does not melt but deforms like a plastic around the obstacle.

Regelation slip

Pressure increases on the up-ice side of small obstacles of <1m. If the pressure is sufficient to cause melting, the meltwater will flow around or over the obstacle and refreeze on the other side where the pressure is lower. Regelation slip is more common when ice flows over smaller obstacles, as less pressure is needed.

| Figure 2.6 | **(a) Basal creep and (b) regelation slip** |

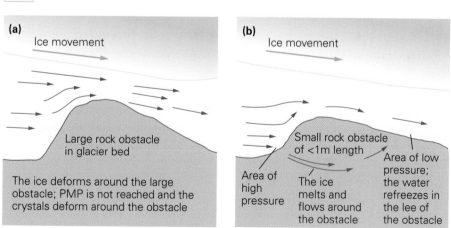

Subglacial deformation

Subglacial deformation is ice movement where the weight of the ice, and therefore pressure, causes the subglacial material to deform and move the overlying ice. This takes place in warm-based ice masses where there is a film of water at the base. Water pressure in the pores between the sediment particles is increased due to pressure melting and this water reduces the cohesiveness of the particles, which are able to move relatively easily.

Activity 2

Using the information in Table 2.2, draw scattergraphs to show the relationship between ice thickness and the amount (%) of movement caused by basal slippage and internal flow. Explain the relationships you find.

Table 2.2 Contribution of different types of flow to movement of glaciers

Glacier	Country	% basal slippage	% internal flow	Ice thickness (m)
Aletsch	Switzerland	50	50	137
Tuyuksu	USSR	65	35	52
Salmon	Canada	45	55	495
Upper Athabasca	Canada	75	25	322
Lower Athabasca	Canada	10	90	209
Blue	USA	9	91	26
Skautbreen	Norway	9	91	50
Meserve	Antarctica	0	100	80

Speed of ice movement

Two further factors affect the speed of glacial movement:
- Ice speed is affected by the gradient of the glacier surface and therefore the shear stress applied to the ice crystals. A steeper gradient will lead to faster movement, with warm-based glaciers moving faster than cold-based glaciers.
- The thickness of the ice also affects the speed of the glacier. The thicker the ice, the more potential there is for internal deformation due to increased stress, and the greater the possibility of the pressure melting point being achieved. Therefore, thicker ice flows faster than thinner ice.

One factor that slows down ice movement, irrespective of the basal thermal regime, is the amount of energy used in overcoming friction between the glacier bed and the sides of the valley. Wherever the ice is in contact with the surrounding bedrock, energy is used to overcome friction. Therefore, across the surface of a glacier, the fastest movement is likely to be found in the centre of the ice, with slower velocities at the sides.

Response of glaciers to flow speed

Evidence of the variable speed of glacier flow is seen in the form of **crevasses**, where flow velocities are greatest. Several types of crevasse can develop as the ice moves: marginal, transverse, radial and longitudinal (Figure 2.7).

Marginal crevasses

The drag exerted on the ice by the sides of the valley causes the margins of the ice to move more slowly. As the ice is more brittle on the surface, it tends to fracture and crevasses develop at right angles to the margin of the glacier.

Transverse crevasses

Transverse crevasses form as the valley gradient becomes steeper and the ice is stretched over the steeper profile (known as **extending flow**). This thins the ice and forces it to fracture. Transverse crevasses can close when the valley gradient is reduced and **compressive flow** occurs.

Radial crevasses

Radial crevasses occur where the snout of the glacier spreads out. As it does so, the ice fractures.

Longitudinal crevasses

Longitudinal crevasses are similar to radial crevasses except they develop where a valley widens and the ice spreads out.

Figure 2.7 **Types of crevasse: (a) marginal crevasses (plan view), (b) transverse crevasses (side view), (c) radial and longitudinal crevasses**

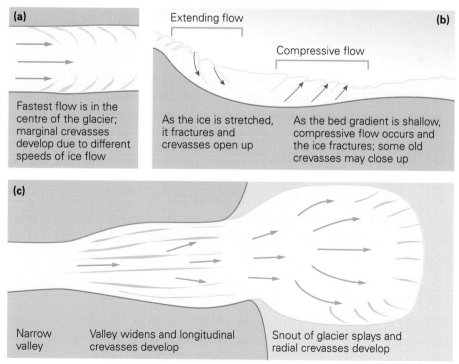

(a)

Fastest flow is in the centre of the glacier; marginal crevasses develop due to different speeds of ice flow

Extending flow

Compressive flow

(b)

As the ice is stretched, it fractures and crevasses open up

As the bed gradient is shallow, compressive flow occurs and the ice fractures; some old crevasses may close up

(c)

Narrow valley

Valley widens and longitudinal crevasses develop

Snout of glacier splays and radial crevasses develop

Rapid glacial movement

There are two types of rapid glacial movement: surges and ice streams.

Glacial surges

Surges are periods of rapid movement as the glacier snout advances up to a thousand times faster than normal. In some cases, glacial surges have reached speeds of 75 m per day (e.g. Black Rapids glacier, Alaska in 1937). Surges are considered to be the result of a change in the flow pattern of subglacial meltwater:

- Water builds up underneath the glacier during a phase of normal glacier flow and there is an increase and thickening of ice in the accumulation area (known as the **reservoir area**).
- During winter, the subglacial meltwater tunnels are closed (due to little, if any, pressure melting) and there is increased accumulation of ice.
- The mass of ice becomes such that in summer, due to the weight of ice, the subglacial tunnels do not open.
- Pressure melting point is reached and the subglacial water separates the basal ice from its bed. This means there is more lubrication and the overlying ice flows more readily.
- Abundant meltwater also increases the pore-water pressure of the subglacial sediment, which adds to rapid movement.
- Once the surge has occurred, the glacier resumes normal flow and enters a period of quiescence. This whole cycle can occur as frequently as every 10–20 years.

Surge events are frequent in the Alaskan glaciers. Variegated glacier surges approximately every 20 years. Its last documented surge was in 1982–83 when it attained a speed of 65 m per day. Hubbard glacier in Alaska last surged in 1986. This surge event was fed by a surge in Valerie glacier, a tributary glacier. Snowfall of 8.5 m in the St Elias Mountains increased accumulation on Valerie glacier and the surge wave passed on to Hubbard glacier.

During the surge event, Hubbard glacier reached a maximum speed of 34 m per day, contrasting with its average speed of 15 cm per day. This surge event proved to be almost catastrophic for the local village of Yakutat. The village is accessed via Russell Fjord. However, the glacier advanced so far that it crossed the fjord and temporarily blocked access to Yakutat. This natural ice dam created a temporary lake that survived for 131 days.

Ice streams

An **ice stream** occurs within an ice sheet and is defined as 'an artery of fast flow surrounded by slower flow'. Sometimes ice streams can contribute to the rapid disintegration of an ice sheet through catastrophic drainage. Ice streams are capable of rapidly reducing the mass of ice in the interior of an ice sheet, which can lead to the disintegration of the ice sheet. Indeed, there is increasing evidence of ice streams contributing to the disintegration of palaeo-ice sheets, such as the Laurentide and the British ice sheets of the last glaciation.

The rapid flow of ice streams occurs when subglacial deformation increases the rate of movement of a section of the ice sheet. Ice streams often have a well-defined morphology, with a wide source area fed by more than one glacier. The output of ice streams is often icebergs.

Ice streams that drain Antarctica account for only 10% of the total ice volume of the continent. They can be up to 50 km wide, 2000 m thick, hundreds of km long and can reach speeds of up to 1000 m per year.

The Jakobshavns Isbrae ice stream in Greenland drains approximately 6.5% of the Greenland ice sheet. During the period 1997–2003, this ice stream accelerated and almost doubled its speed (Table 2.3). The increased velocity has led to the rapid thinning of the Greenland ice sheet, with an annual reduction of 15 m of thickness each year since 1997.

| Table 2.3 | Ice stream episodes of the Jakobshavns Isbrae ice stream in Greenland |

Year	Velocity (m per year)
1985	6700 m per year
1992	5700 m per year
2000	9400 m per year
2003	12 600 m per year

Table 2.4 shows the dimensions and speed of four ice streams in Antarctica and Greenland.

| Table 2.4 | Dimension and speed of four ice streams in Antarctica and Greenland |

Ice stream	Length (km)	Width (km)	Thickness (m)	Velocity (m per year)
Ice stream A	>200	~50	~1000	215–254
Ice stream E	~320	75–100	975–1091	400–550
Pine Island glacier	200	26	1564	1300–2600
Jakobshavns Isbrae	70–80	~6	2500	800–7000

3 Glacial erosion

Processes of glacial erosion

Glaciers are responsible for some of the most impressive erosional landforms on the planet. How are these landforms created? Essentially, they are the result of two main erosional processes: abrasion and plucking.

Abrasion

Abrasion occurs where glacier ice is in contact with the valley bottom and sides. Ice at the bottom of the glacier carries debris with it and this, in turn, scratches or abrades the bedrock surface. Clear ice alone is not able to scratch the surface of the rock and so ammunition is needed, in this case debris. Abrasion is restricted to warm-based glaciers, and the amount and rate of glacial abrasion is determined by several factors, as shown in Figure 3.1.

Figure 3.1 **Factors that affect the amount and rate of glacial abrasion**

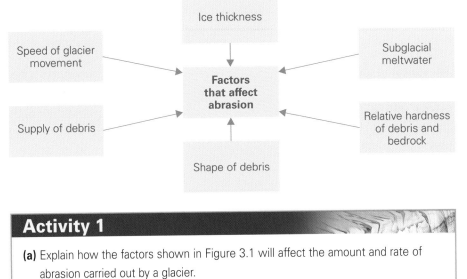

Ice thickness

Speed of glacier movement

Subglacial meltwater

Factors that affect abrasion

Supply of debris

Relative hardness of debris and bedrock

Shape of debris

Activity 1

(a) Explain how the factors shown in Figure 3.1 will affect the amount and rate of abrasion carried out by a glacier.

(b) Find examples of different glaciers to illustrate these various factors.

Abrasion is evident in the formation of **glacial striations** or **striae** (scratches on bedrock), examples of which are shown in Figure 3.2.

Figure 3.2 | **Striations on the Ungava Peninsula, northern Quebec, Canada. The presence of two directions of striations on the rock surface suggests that this area has experienced two phases of ice flow.**

The grinding down of the bedrock and debris as the ice moves produces a fine powder known as **glacial flour**. This is eventually washed from beneath the glacier and is transported into meltwater streams. Glacial flour carried in suspension is responsible for the distinctive blue and green colour of many meltwater streams.

Plucking

Plucking is the second process of erosion and is considered by many scientists to be more effective than abrasion. It is also known as **joint block removal**. Again this process occurs beneath the ice and is dependent on the movement of the glacier. Plucking requires the bedrock to be jointed and therefore weakened. Ice freezes onto the bedrock and then the blocks of bedrock become incorporated into the sole of the glacier.

How does this happen? Melting (pressure melting point) and refreezing need to occur and this is most likely to happen where there is an obstacle obstructing the flow path of the glacier. The process of regelation occurs: ice melts due to increased pressure and then refreezes onto the rock where the pressure is reduced.

Sediment source

Abrasion needs sediment at the base of the glacier, while plucking occurs where rocks are well jointed. So how do rocks beneath a glacier become jointed and what is the source of sediment for abrasion?

It is thought that subglacial rocks become jointed prior to glaciation. During this time, **periglacial** conditions (see Chapter 6) prevail. Periglacial areas are found in front of current glaciers and ice sheets and coincide with arctic tundra vegetation. In periglacial areas, the weathering process of **freeze–thaw** is important (Figure 3.3). Water enters the rock through joints, faults and pores. As temperatures fall, the water freezes and there is a volumetric increase of 9%. This exerts great pressure on the surrounding rocks.

Observations suggest that the greatest pressure is felt at −22°C, when the pressure can reach 2100 kg cm^{-1}. However, rocks will break up before this pressure is achieved (Figure 3.4). For example, granite has a tensile strength of 70 kg cm^{-1}. However, the pressure exerted on the rock as the ice expands is thought to be no more than 14 kg cm^{-1}. Of course, for the ongoing cycle to continue, melting must occur and there needs to be a temperature fluctuation above and below zero.

| Figure 3.3 | **Freeze–thaw cycles; more than one cycle is needed for freeze–thaw to be effective** |

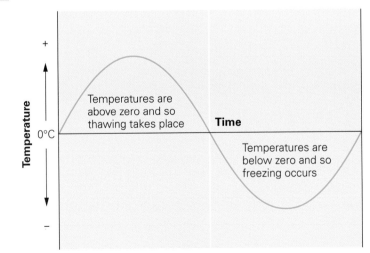

Figure 3.4 Frost-shattered rock at Bowscale Fell in the Lake District

J. Knight

Furthermore, as a glacier erodes a valley bottom, it removes bedrock and replaces it with ice, which is lighter and just one-third the density of bedrock. This causes the bedrock to expand or dilate, a process known as **dilatation**. As the rock expands, it fractures and joints are created. Similarly, if ice melts at the snout of a glacier (effectively, weight is removed) then dilatation occurs. The result is a network of fractures in the rock that are parallel to the surface.

Landforms and landscapes

There are several characteristic landforms and landscapes that are created by erosional processes. The principal ones are: knocks and lochans, glacial troughs, corries, roches moutonnées and striations.

Knock and lochan landscapes

Ice sheets such as the Laurentide ice sheet and the British ice sheet during the Devensian glacial eroded large areas so extensively that bare rock was left exposed. This type of erosion is termed **areal scouring** and the landscape that results consists of thousands of low-lying hills, called **knocks,** and intervening lakes, called **lochans** (Figures 3.5 and 3.6).

Figure 3.5	The knock and lochan topography near Lochinver, northern Scotland

Reproduced by permission of Ordnance Survey on behalf of HMSO. © Crown Copyright (Licence No.100027418)

Figure 3.6	Knock and lochan landscape of Lewisian gneiss, near Loch Assynt, northern Scotland, with Canisp in the background

M. Raw

Knock and lochan landscape is pitted with numerous lochans or lakes, which are often aligned in the same direction. This is because the direction of the lakes is influenced by geological features, such as faults and dykes. These weaknesses in the rock can be exploited and eroded by the ice. The lakes therefore appear to have been orientated by the ice flow but in fact the landscape is controlled by the geology. This type of landscape was first named by the geographer Linton in 1963 after studying the landscape in the northwest Highlands of Scotland.

The Ungava Peninsula in northern Quebec was eroded extensively by the Laurentide ice sheet. The landscape here appears to be relatively disorganised, with no obvious large-scale erosional landforms, but the bedrock is extensively abraded and there are many striations present, indicating that ice was key to its formation (Figure 3.7).

| Figure 3.7 | Satellite image of the Ungava Peninsula in northern Quebec — the light areas of the image are glacially scoured bedrock; the dark areas are lakes — another example of knock and lochan |

Image courtesy of the European Space Agency and Sheffield University

Glacial troughs

Former river valleys were occupied by ice during the Devensian glaciation. Glacial erosion widened and deepened these valleys, giving them a U- rather than V-shape. This type of valley is called a **glacial trough.** The precise nature of the valley shape is debatable and is often likened more to a parabola. Only in exceptional circumstances are the valley sides steep enough to be described as a true U-shape.

The valley sides are often formed from preglacial valley spurs that have been eroded by the ice to form **truncated spurs** (Figures 3.8 and 3.9). Today, small rivers often occupy the course of the glacial valley. These rivers are too small to have eroded the valley and are known as **misfit streams**.

Figure 3.8 The glacial trough of the Yosemite valley, USA

Vertical sides of the valley show truncated spurs

Broad wide valley floor

J. Knight

Figure 3.9 The Pasterze glacier, Austria

Truncated spur eroded when the glacier had a greater volume of ice

Glacier trim lines

Lateral moraine

Direction of flow

Supraglacial medial moraine

J. Knight

Activity 2

Figure 3.10 **The Nant Ffrancon valley, Snowdonia, north Wales**

(a) Map of the Nant Ffrancon valley, Snowdonia

Typical U-shaped valley with broad, flat bottom and fairly steep sides

The river Afon Ogwen is a misfit stream

J. Knight

(b) The glacial trough of the Nant Ffrancon valley

Activity 2 (continued)

(a) Using the map in Figure 3.10(a) draw a cross-section of the Nant Ffrancon valley. Describe the cross-section.

(b) Using Figure 3.10(b) draw an annotated sketch to show the main features of the Nant Ffrancon valley.

(c) Describe and explain changes that have occurred to the valley since the end of the last ice age.

Long profile of glaciated valleys

The long profile of a glacial valley is often highly irregular. For example, many glacial troughs have alternating rock bars and depressions, the depressions now being filled with water. Elterwater in Great Langdale fills a depression; there is a rock bar at Skelwith Force and the other end is dammed by a rock barrier at Chapelstile.

What is responsible for these irregularities? There are several possible explanations:

- Narrowing of the valley increases the velocity of ice flow; erosion then increases, resulting in deepening of the valley floor.
- Depressions represent points where tributary glaciers join the main glacier.
- The glacier has exploited and accentuated preglacial irregularities in the valley floor.
- Geological variations exist along the valley floor.
- Compressional and extensional flows have increased erosion in some places.

Fjords

Glacial troughs that end in the sea are often flooded by sea water due to **eustatic sea-level rise**. These drowned glacial troughs are known as fjords and good examples are found on the coasts of Scotland, Alaska, Greenland, New Zealand and Norway. Towards the sea the fjord becomes shallower and there is often a bar at the mouth of the valley. This is caused by the glacier thinning towards its snout and losing its power of erosion. In some places, material is deposited on top of the rock lip.

Corries

Corries (cirques) are common landforms in glaciated upland areas, such as the Lake District, Scottish Highlands, Cairngorms and north Wales. The Lake District, for example, has an estimated 128 corries in total.

Corries are smaller erosional features than glacial troughs. They have a distinctive armchair shape, with a steep headwall and sides, and an over-deepened base. Corries are generally oriented to the north or east where summer insolation values and temperatures are lower during periods of alpine glaciation.

In the British Isles, corries were formed towards the end of the Devensian glaciation, approximately 14000 years BP; the ice disappeared by 10000 years BP. Corries therefore develop in a relatively short period of geological time. This explains their relatively small size. There are several stages in their formation:

- A corrie starts as a **nivation hollow** in the uplands, as snow and ice build up over summer.
- Over time, the ice thickens. Once it is deep enough, it begins to rotate and move under its own weight (Figure 3.11). The greatest pressure is at the bottom of the hollow and this becomes over-deepened by abrasion and plucking.
- A **bergschrund** (a deep crevasse) develops towards the headwall, caused by the tension that develops as the ice extends. Rock debris can then enter the ice through the crevasse and become an erosional tool.
- As the ice moves, plucking occurs on the headwall and this leads to **headward erosion**.

| Figure 3.11 | **Rotational movement of ice in a corrie** |

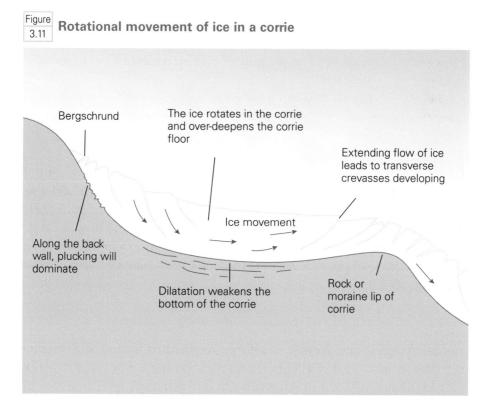

Where two corries are adjacent to each other they are separated by a narrow knife-edged ridge known as an **arête**. Many of the ridge walks in the Lake District are along arêtes, for example Striding Edge near Ullswater and Sharp Edge on Blencathra.

A **pyramidal peak** is a less common landform created when three or more corries experience headward erosion. The most famous example is the Matterhorn in Switzerland. All corries in the UK are now ice free (Figure 3.12), but major mountain ranges, such as the Alps and the Rockies, still support corrie glaciers.

Figure 3.12	**Bowscale Tarn in the Lake District**

Steep back wall

J. Knight

(a) The corrie at Bowscale Tarn is oriented to the north and is developed in Skiddaw slates. During the Devensian glaciation, the corrie is calculated to have covered an area of 450 m × 430 m, with a maximum ice depth of 95 m. The maximum speed of ice flow was 35 m per year.

(b) Bowscale corrie, showing the moraine lip. The lip forms a natural dam for Bowscale Tarn. It is estimated to be 30 m deep, containing 680 000 m^3 of glacial deposits.

J. Knight

Activity 3

Table 3.1 shows the orientation and elevation of corries in Snowdonia National Park.

Table 3.1   Orientation and elevation of corries in Snowdonia National Park

Group 1		
Corrie number	**Orientation (°)**	**Elevation (m)**
1	310	279
2	330	385
3	345	300
4	20	400
5	120	285
6	140	400
7	70	385
8	340	335
9	345	335
10	155	420
11	85	370
12	40	285
13	10	285
14	95	335

Activity 3 (continued)

Group 2

Corrie number	Orientation (°)	Elevation (m)
1	10	950
2	30	1000
3	240	1400
4	310	1900
5	20	1250
6	320	1700
7	85	2000
8	40	2500
9	0	2250
10	180	1150
11	80	2800
12	70	1400
13	190	1050
14	80	1100
15	175	1250

Group 3

Corrie number	Orientation (°)	Elevation (m)
1	330	1750
2	340	2000
3	30	1700
4	60	1800
5	45	1500
6	55	1500
7	75	1550
8	50	1750
9	80	2100
10	50	1250
11	10	2500
12	15	2400
13	10	1800
14	35	1500
15	45	1800
16	55	1250
17	85	1250

Activity 3 (continued)

Group 4		
Corrie number	Orientation (°)	Elevation (m)
1	350	1600
2	355	2500
3	85	1750
4	45	2750
5	80	2050
6	170	1800
7	45	1600
8	110	1750
9	5	1400
10	120	2200

Figure 3.13 shows the location of the corries.

Figure 3.13 Location of four groups of corries in Snowdonia National Park

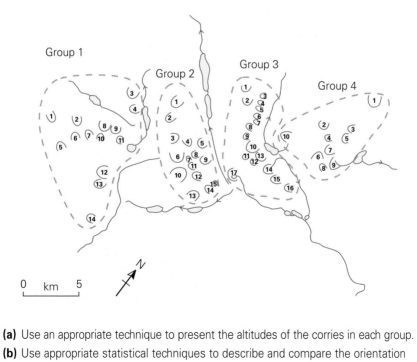

(a) Use an appropriate technique to present the altitudes of the corries in each group.

(b) Use appropriate statistical techniques to describe and compare the orientation of the four groups of corries.

Activity 3 (continued)

(c) Use an appropriate technique to present the orientation of the four groups of corries.

(d) Comment on your results.

Roches moutonnées

Roches moutonnées are outcrops of bedrock that have been eroded and stream-lined in the direction of ice flow. They are often known as 'whalebacks' because of their appearance. Roches moutonnées are formed underneath the ice and are eroded by plucking and abrasion as the ice passes over them. At the down-ice end of the roche moutonnée, regelation occurs.

Striations

Striations (or striae) are abrasion marks on the surface of rocks. They often cut across faults and joints in rocks. Although not strictly a landform, striations are the result of the process of abrasion. This means that the orientation of striations show us the orientation of ice flow.

Activity 4

The striation orientations shown in Table 3.2 were collected from northern Quebec.

Table 3.2   Orientation and frequency of striations at a site in northern Quebec

Orientation (°)	Frequency	Orientation (°)	Frequency
360	3	28	3
2	3	29	3
10	3	32	7
12	3	35	3
18	7	38	3
19	7	40	2
22	3	42	3
25	7	51	3
26	8	56	2
27	7	60	2

(a) Plot the data using a suitable method of data presentation.

(b) Use descriptive statistics to describe the data set shown.

(c) In which orientation did the ice flow?

4 Glacial debris

Transportation and deposition

As a glacier moves it **transports** vast amounts of debris. Much of this material is used to erode the landscape and is then **deposited**.

Transportation

Glaciers carry material in several ways:
- supraglacial transportation
- englacial transportation
- subglacial transportation

Irrespective of the zone in which material is transported, it generally remains angular and unsorted.

Supraglacial transportation

This is where frost-shattered and avalanche material falls onto the glacier surface from valley sides. This material will be very angular and unsorted as it has not undergone abrasion.

Englacial transportation

As the ice moves, it deforms around obstacles, leading to the formation of crevasses. Slope debris falling into these crevasses becomes trapped in the glacial ice and is moved within the glacier.

Subglacial transportation

Some of the englacial material will eventually make its way to the glacier base and from there it is transported. Material in this zone can also be sourced by subglacial plucking.

Deposition

Glaciers deposit sediment in front of or adjacent to the ice, creating distinctive landforms. Glaciers also deposit material beneath the ice. These deposits often cover vast areas of land.

Deposited material has several names:

- **glacial till**
- **boulder clay**
- glacial drift
- diamicton
- **moraine**

The deposited material is a matrix of fines, such as sand and clay, mixed with larger **clasts** and boulders. There is no sorting, which means that there is no order in which different-sized particles are deposited (it is unstratified), and the clasts tend to be angular.

Activity 1

Explain why the material transported by glaciers remains unsorted.

Glacial deposition takes the form of **moraine ridges,** which can run parallel to the snout or perpendicular to the glacier, or deposition can be **subglacial.** Both valley glaciers and ice sheets deposit material but the scale of deposition differs. For example, the Laurentide ice sheet had a distance of up to 600 km between the ice margin and the terminal moraine as it retreated, with several smaller recessional moraines deposited, located behind the terminal moraine. In contrast, a valley glacier tends to have a depositional area between the glacier margin and terminal moraine of less than 10 km.

Glacial deposition landforms

Landforms at the snout

Depositional landforms owe their character to:

- the amount of sediment that the glacier is transporting
- the glacier's mass balance state; for many depositional landforms the glacier needs to be experiencing a negative mass balance

Several different types of moraine ridge can be formed by glacial deposition:

- terminal moraines
- recessional moraines
- push moraines

Terminal moraines

A **terminal moraine** is a ridge of moraine deposited in front of a glacier. A glacier has only one terminal moraine, and it marks the greatest extent of the glacier. Three stages are important in the formation of terminal moraines:

- The glacier has a positive mass balance, which causes the ice to advance.
- Boulder clay is pushed along by the snout of the glacier and forms a pile.
- The glacier retreats due to a negative mass balance, leaving the pile of unsorted angular glacial debris as a ridge.

Terminal moraines develop parallel to the ice margin and are typically between 30 m and 60 m high. The Franz Josef glacier in New Zealand had the highest terminal moraine on record: it stood 430 m high.

Activity 2

Explore the factors that control the height of a terminal moraine.

Recessional moraines

A **recessional moraine** also forms parallel to the glacier snout, but represents episodes in the retreat of the glacier in response to a negative mass balance. As the snout retreats, it sometimes temporarily stands still — long enough for the debris that is being transported to be deposited. It is important to remember that even though the snout of the glacier is at a standstill, the ice continues to flow. As the inputs are equal to the outputs, material can still be added to the moraine ridge.

Proglacial areas (defined as the area of land in front of a glacier) can have more than one recessional moraine. They will be positioned between the glacier snout and the terminal moraine. The sediments in these moraine ridges tend to be unsorted and angular.

Push moraines

Push moraines are also formed parallel to the ice margin. They are small ridges that result from small short-lived winter re-advances. Therefore, while terminal and recessional moraines are formed in retreat and are structureless, push moraines are formed during an advance. As a result, the nature of the sediments tends to be slightly different.

Push moraines often reveal annual advances of a glacier and include glacio-fluvial sediment of more rounded clasts in small ridges. It is not uncommon for push moraines to be obliterated from one winter to the next because of summer melting. They can also disappear because of an increase in snow input into the system, which causes the snout to advance.

Examples of push moraines are found in front of the glaciers on Axel Heiberg Island in the Canadian High Arctic.

Moraines at the side of glaciers

Two types of moraine form at the side of glaciers:

- lateral moraines
- medial moraines

Lateral moraines

Lateral moraines are formed on top and to the side of valley glaciers and run parallel to the direction of ice flow (Figure 4.1). A lot of material is input from **frost shattering** and **mass movement** on the exposed valley walls above the glacier, and this material tends to be angular. Some of the lateral moraine will rest on the ice and some will rest on the substrate. If the glacier shrinks, some lateral moraines can be left perched above the former ice level. Material is incorporated into the ice if the glacier is heavily crevassed.

| Figure 4.1 | **Lake Louise lateral moraine, Canadian Rockies** |

Lateral moraine

J. Knight

Medial moraines

Medial moraines develop at the confluence of two valley glaciers which are both carrying lateral moraines (Figure 4.2). Again the main source of material is from frost shattering and mass movement on valley slopes. The medial moraine ridge tends to be slightly elevated above the glacier as it insulates the ice beneath it and therefore reduces ablation.

Survival of medial moraines depends on the supply of sediment. As the moraine extends towards the snout, it becomes a less well-defined ridge and more of a till sheet on the ice surface.

Figure 4.2 **Medial and lateral moraines on the Pasterze glacier, Austria**

Lateral moraine caused by glacier shrinkage

Medial moraine

J. Knight

Activity 3

Draw a diagram to show the different locations of moraine ridges in relation to the glacier snout.

Activity 4

(a) Use the following websites to find examples of moraine ridges:
   www.uwsp.edu/geo/faculty/lemke/alpine_glacial_glossary/glossary.html
   www.swisseduc.ch/glaciers/earth_icy_planet/glaciers06-en.html
(b) Complete a copy of Table 4.1 with information about different types of moraine.

Activity 4 (continued)

Table 4.1 Different types of moraine

Name	Type of moraine	Location	Description

Moraines deposited in water

Two further types of moraine are found in areas that have been covered by ice sheets: **rogen** and **De Geer** moraines. Unlike the moraines described in the previous section, these have been deposited in water.

Rogen moraines

Rogen moraines are irregular landforms 10–20 m high, 50–1000 m wide and 1–2 km long. They are slightly arcuate in their appearance and concave in the up-ice direction. Rogen moraines are separated by lakes, and they often mark the transition to drumlins formed beneath extensive ice sheets, such as the Laurentide and European ice sheets. It is thought that rogen moraines were deposited in water at the ice margin during recession of the ice. Extensive rogen moraines are found in northern Quebec, the Northwest Territories (Figure 4.3) and Alberta (Canada), formed as the Laurentide ice sheet retreated.

Figure 4.3 **Aerial photo of rogen moraines at Boyd Lake, Northwest Territories, Canada**

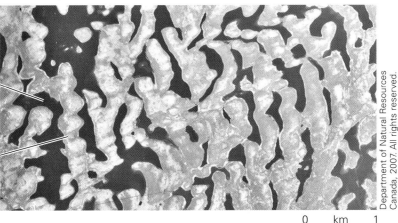

The dark areas are lakes that separate the rogen moraines

The elongated 'islands' are the rogen moraines

0   km   1

De Geer moraines

De Geer moraines are also deposited transverse to the direction of ice flow (parallel to the margin). They occur some distance behind ice margins that calve into a lake. They are a succession of discrete narrow ridges that rarely exceed 15 m high, separated by up to 300 m (Figure 4.4). They are thought to form where an ice sheet enters a lake or the sea. At the land–water margin, the bed of the ice sheet becomes decoupled from the bedrock and debris accumulates between the bed and the ice. However, while this is the most likely explanation, their formation is not well understood. Extensive outcrops are found on the Ungava Peninsula and around Hudson Bay, formed as the Laurentide ice sheet decayed.

| Figure 4.4 | **De Geer moraines at Generator Lake, Baffin Island, northern Canada** |

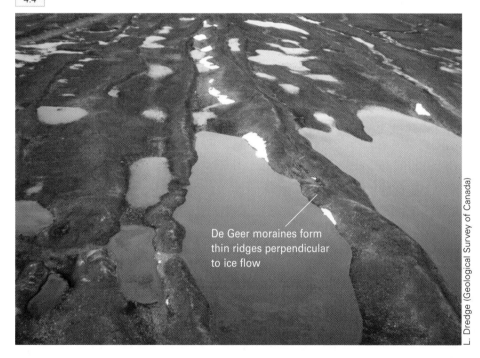

De Geer moraines form thin ridges perpendicular to ice flow

L. Dredge (Geological Survey of Canada)

Subglacial deposition

Ice sheets are able to deposit **subglacial till** across extremely large areas, so depositional features are not only moraine ridges but also extensive plains, such as those found in East Anglia, UK. Not only are these moraines more extensive than those deposited at the ice margin and at the sides of glaciers, but they also tend to show more sorting of the particles, which are often aligned to the direction of ice flow.

Lodgement till

As a glacier begins to lose its erosive power, it deposits material subglacially. In warm-based glaciers, pressure melting encourages deposition, especially where the bed is irregular. Till is often plastered against any obstacle. Some of the till is squeezed into gaps in the bed, creating a smooth surface.

Because pressure is exerted on the till as it is forced against obstacles, so the clasts in the till take on a preferred orientation and are lodged at an angle of about 45°. The rate of subglacial lodgement deposition is considered to be about 6 m per century (Figure 4.5). **Lodgement till** is responsible in part for the formation of the subglacial features known as drumlins.

Figure 4.5 **Boulder clay at Criccieth, north Wales**

Height of cliff face approximately 3.5 m

There is an enormous range in the size of the clasts deposited by the ice

The finest material is known as the till matrix; here it is mostly clay

Meltout till

Meltout till is commonly found at ice margins where ice is stagnant and ablation is occurring. It can also happen subglacially where there is a large geothermal heat flux that melts the base of the glacier. As the ice melts, material transported by the ice is deposited.

Activity 5

The data in Table 4.2 were collected from two different till deposits at Blea Water Tarn in the Lake District.

Table 4.2 Till deposit samples from Blea Water Tarn

| Clast number | Till sample 1 | | Till sample 2 | |
	Orientation (°)	Length of A axis (cm)	Orientation (°)	Length of A axis (cm)
1	100	4.8	162	3.9
2	70	6.9	17	3.5
3	120	4.0	140	5.9
4	120	6.5	72	4.0
5	100	7.0	61	31.0
6	166	5.6	100	11.4
7	104	6.0	100	7.0
8	152	9.8	149	7.0
9	120	12.0	58	12.2
10	121	15.5	18	4.8
11	100	6.0	89	10.0
12	80	7.0	156	9.6
13	170	4.5	120	11.0
14	232	10.5	139	3.6
15	225	7.5	14	10.9
16	227	6.0	32	14.0
17	85	11.5	32	7.2
18	70	22.0	126	8.0
19	54	20.5	121	7.0
20	95	5.0	51	7.6

(a) Use an appropriate technique to present the orientation of the two samples to see if they are similar or different.

(b) Construct two histograms to show the size of the clasts.

Activity 5 (continued)

(c) Conduct a statistical test of difference, such as the Mann Whitney *U* test or the chi-squared test, to see if there is a statistical difference between the orientation of the two samples.

(d) Are these two samples from the same till or are they from different tills? How can you tell?

Subglacial deposition landforms

The most notable landforms deposited by ice are **drumlins** and **glacial lineations**. Both of these form parallel to the direction of ice flow and result from the deformation of the subglacial till.

Drumlins

Drumlins are elongated hills that are aligned parallel to the direction of ice flow (Figures 4.6 and 4.7). They can occur singly or in swarms of over a hundred. Where they occur in swarms, they are also known as 'basket and egg topography'. The most extensive drumlin swarm known is found in New York state and was formed by the Laurentide ice sheet. There the field is 225 km × 56 km and it contains over 10 000 drumlin forms. The most extensive drumlin swarms in the British Isles are in Scotland, the northwest of England and western Ireland.

Figure 4.6 **Drumlin swarm, showing the stoss and lee ends of drumlins and their orientation**

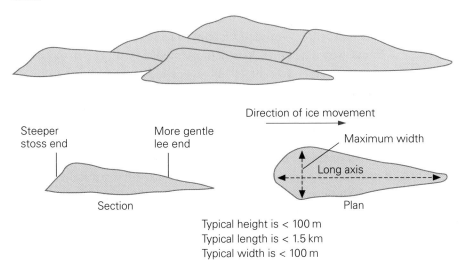

Direction of ice movement

Steeper stoss end

More gentle lee end

Maximum width

Long axis

Section

Plan

Typical height is < 100 m
Typical length is < 1.5 km
Typical width is < 100 m

Figure 4.7 **Drumlin scenery in the Rawthey/Lune valley, northwest England**

The arrows show the orientation of the
long axes of some of the drumlins

The explanation behind drumlin formation intrigued glacial geomorphologists
for many years. This is because drumlins are formed subglacially and are only
revealed once the ice has disappeared. Therefore it is difficult to observe their
formation directly. In recent years, there have been two popular theories
explaining their formation.

- **Subglacial deformation of till**. This theory was proposed by Boulton (1987),
  who suggested that as ice overroded the subglacial till, it moulded the till into
  elongated forms, which had their long axis oriented in the direction of ice flow.
- **Active basal meltwater**. Shaw et al. (1989) suggested that subglacial meltwater
  carved cavities into the base of the ice and these cavities were in-filled with
  subglacial sediments as the meltwater ceased to flow. Large drumlin fields
  were explained as being the result of catastrophic meltwater floods.

Till fabric analysis suggests that subglacial deformation is more likely, although
the fabric of drumlins is highly variable. Drumlins are usually composed of till
and some have the long axis of clasts orientated in the direction of ice flow while
others exhibit a less clear fabric. Some drumlins have a rock core around which
deposition has occurred, while others contain different materials that may
encourage deposition.

Glacial lineations

Both drumlins and glacial lineations provide important evidence of the direction and possibly the speed of ice flow in the past. Figure 4.8 shows a satellite image of glacial lineations in northern Quebec. Of even greater significance are situations where more than one direction of glacial lineation has been super-imposed on another, suggesting two separate phases of ice flow.

Figure 4.8 Satellite image of glacial lineations in northern Quebec, Canada

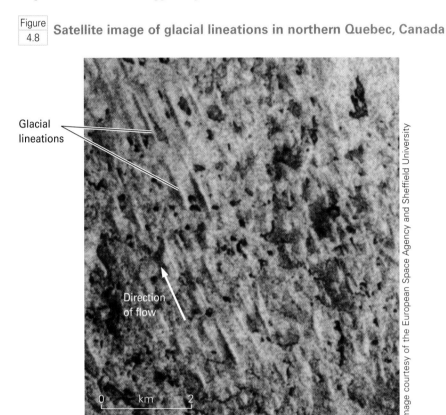

Glacial lineations

Direction of flow

0 km 2

Image courtesy of the European Space Agency and Sheffield University

The chronology of such ice-flow events is important for industries such as mineral extraction as the ice will have moved minerals from one place to another. Clearly if the sequence of ice movement is known, it is possible to trace the location of minerals such as sand and gravel.

5 Landforms developed by meltwater

Meltwater

Meltwater is one output of the glacial system. It is generated in three ways:

- ablation due to incoming short-wave solar radiation in the summer melting season — a surface (supraglacial) source of meltwater
- subglacial melting when the ice reaches pressure melting point — a subglacial source of meltwater
- precipitation — another surface source of water

Surface sources of meltwater reach the base of the glacier by flowing down crevasses or **moulins**. Then they either flow through the ice (englacial meltwater) or reach the bottom of the glacier. From the subglacial area, meltwater flows out from beneath the glacier into the proglacial area. This means that distinctive glaciofluvial landforms develop both beneath the ice, exposed during deglaciation, and in front of the ice margin.

Glaciofluvial sediments

Sediments transported by meltwater are distinct from the materials transported by ice. So too are the landforms. Glaciofluvial sediments differ from true glacial deposits in two main ways:

- Meltwater sediments are more rounded than true glacial sediments due to the erosional processes of attrition and abrasion, which smooth particles that were originally angular.
- Meltwater deposits have been sorted, with larger particles deposited first. Sorting is determined by the changing **competency** and **capacity** of meltwater streams. During the warmer summer months there is a daily rhythm to the discharge of meltwater.

Sedimentation sequence

The rhythm of deposition of glaciofluvial sediments has a daily and an annual pulse. During the summer months, there is an increase in the discharge of subglacial and proglacial streams due to heating of the atmosphere and the resultant melting of glacier ice. Proglacial and subglacial streams reach peak discharge during the late afternoon. By sunset, discharges have already decreased in response to falling temperatures. As discharge declines, larger sediments are deposited first and finer sediments later. This gives rise to distinct layers and sequences where the sediments get finer towards the top of the bed.

In proglacial lakes, coarser sediments are deposited in the summer months when there is more energy available for river transport, and finer materials are deposited towards the end of the melt season in the autumn as river levels fall. Over winter, the finest materials settle out. This gives rise to **varves** or layers of sorted sediment (Figure 5.1).

| Figure 5.1 | Typical meltwater sediment in the Yosemite valley, USA. The coarsest materials are at the base of the sequence and the finer sediments are higher up. |

Fine sediment

Coarse sediment

J. Knight

Glaciofluvial landforms

Eskers

Eskers are glaciofluvial landforms that form within or beneath warm-based glaciers. They are made up of sinuous ridges composed of sands and gravels. Eskers are aligned parallel to local striations — in the direction of former ice flow — and can reach a height of 200 m, a width of 3 km and up to 100 km in length.

Eskers form either in meltwater tunnels that develop at the base of the ice (e.g. Rothlisberger and Nye channels) or in meltwater tunnels that are completely surrounded by ice (englacial tunnels). In both cases, four conditions are necessary for their formation:

- running water
- high amounts of transported sediment
- a change in the velocity of flow, resulting in deposition
- retreat of the ice margin to expose the esker

Beneath a glacier or ice sheet, meltwater flows through a network of tunnels, often under hydrostatic pressure. The pressure is so great in places that the water is even forced up gentle gradients. When the discharge is high, a lot of sediment is transported. However, when the discharge falls at the end of the summer melt season, sediment is quickly deposited. Then subglacial or englacial tunnels fill with sediment, which is exposed when the ice margin retreats.

The sediments that form eskers beneath the ice start off being angular as they are derived from the glacier but become rounded and smooth due to fluvial erosional processes. Essentially, eskers are the debris that once filled ancient subglacial tunnels. They reveal major routes followed by meltwater (Figure 5.2).

| Figure 5.2 | **The formation of subglacial eskers** |

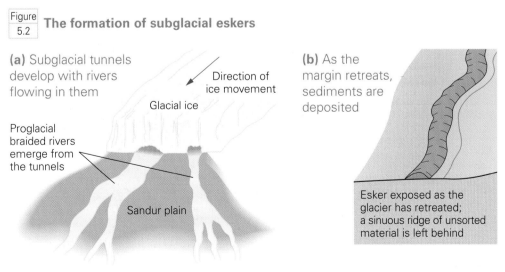

(a) Subglacial tunnels develop with rivers flowing in them

Direction of ice movement

Glacial ice

Proglacial braided rivers emerge from the tunnels

Sandur plain

(b) As the margin retreats, sediments are deposited

Esker exposed as the glacier has retreated; a sinuous ridge of unsorted material is left behind

The eskers present over large areas of Canada and the USA formed as the Laurentide ice sheet retreated. North Dakota is littered with eskers of varying length and width. They provide clues to the way the ice sheet disintegrated as the climate warmed. The Dahlen esker in Grand Forks County, North Dakota, USA, is particularly well preserved; it is approximately 6.4 km long, 122 m wide and reaches elevations of 15–24 m.

One of the largest eskers in England is Blakeney esker in north Norfolk. It extends southeast from Blakeney to Wiveton Downs, northwest of Glandford (Figure 5.3). Its total length is 3.5 km and it rises 20 m above the surrounding land.

Figure 5.3   **Blakeney esker**

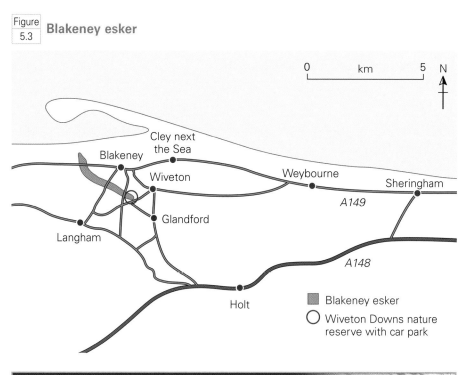

Activity 1

Use the website www.bgs.ac.uk/blakeney/BlakeneyEsker.htm to complete the following tasks:

(a) Describe the nature of the sediments found in Blakeney esker.

(b) Briefly describe the formation of the esker.

(c) In what way is the formation of Nye channels beneath a glacier an example of positive feedback?

(d) What use has Blakeney esker been to humans?

Kames

Kame is a word meaning 'rounded hill'. Like eskers, kames are meltwater deposits formed during deglaciation. Kame sediments are usually bedded and comprise sorted sands and gravels. Occasionally they have sloping beds, suggesting deposition in a pond or lake. However, their formation is not fully understood. At least two theories have been suggested to explain their formation.

In the first theory, meltwater streams with a heavy sediment load flow into a proglacial lake. The material is deposited in deltaic forms, with one side of the deposit remaining in contact with the ice. During deglaciation, the ice melts and its support is removed, leading to slumping of the sediments (Figure 5.4).

Figure 5.4 **The first theory of kame formation**

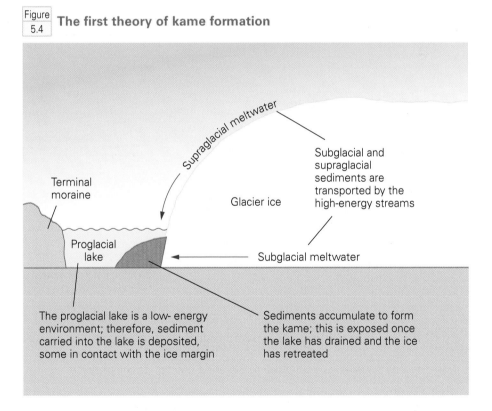

A second theory was put forward by Holmes in 1947. He suggested that pools or ponds develop on the surface or within stagnating ice as meltwater flows into depressions. The sediment builds up into a mound and as the ice around the deposit melts, the kames are lowered and deposited at the base of the ice (Figure 5.5).

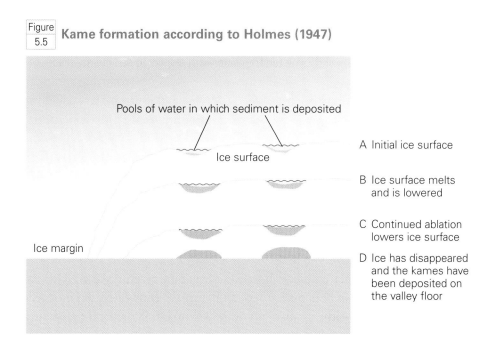

Figure 5.5 Kame formation according to Holmes (1947)

Pools of water in which sediment is deposited

Ice surface

Ice margin

A Initial ice surface

B Ice surface melts and is lowered

C Continued ablation lowers ice surface

D Ice has disappeared and the kames have been deposited on the valley floor

Good examples of kames formed at the end of the last glaciation are found in the Cairngorms, Scotland. Kames are also found in north Norfolk near to the Blakeney esker. They are between 20 m and 400 m in diameter and up to 20 m high.

Activity 2

What are the strengths and weaknesses of the two theories of kame formation?

Kame terraces

Kame terraces are elongated mounds of sorted sand and gravel. They are formed from sediment deposited by meltwater in contact with the ice margin. However, kame terraces differ from kames because they form at the sides of glaciers.

Their development starts when heat from the valley sides melts the ice that is in contact with it. During the summer, meltwater runs along the edge of the glacier and can form marginal lakes. At times of low meltwater flow, material is deposited in layers, fining upwards. In this way the marginal lakes are infilled. The sediment is well bedded and comprises sorted sand and gravel. This material remains in contact with the ice until the glacier wastes downwards, leaving a terrace stranded at the side of the valley.

Activity 3

How do lateral moraines and kame terraces differ with respect to sediments, shape and form?

Activity 4

Look at the following website and describe the formation and characteristics of the kame terraces found in the Cairngorms:

www.fettes.com/cairngorms/kame%20terrace.htm

Sandur plains

The area in front of a glacier or ice sheet is known as the **sandur plain,** proglacial plain or **outwash plain**. It is made up of a mixture of sediments from morainic deposits, which are clearly angular and derived from glacial transportation, and materials that have been transported and deposited by meltwater so are slightly more rounded. The sediments tend to be reworked due to the changing discharge of proglacial streams and so they become reasonably well mixed, although vertically there is some bedding (Figure 5.6).

Figure 5.6 **The proglacial area and the ice marginal environment of the Pasterze glacier, Austria**

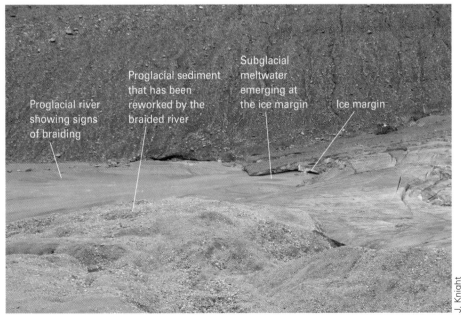

J. Knight

Kettle holes

Kettle holes are hollows found in sandur plains or vast till plains, such as those that cover Indiana in North America. They are responsible for giving a sandur plain its dimpled appearance (Figure 5.7). As an ice sheet retreats, some blocks of ice may become detached. These blocks are known as **dead ice**. Often dead ice becomes embedded into the sandur plain and is covered with sediment. As the climate warms, the dead ice melts and leaves a depression in the surface. Meanwhile, surrounding sediment is made unstable and collapses into the pit. In some places kettle holes are now lakes or small circular patches of boggy ground (e.g. in Scotland).

Figure 5.7 **A kettle hole in the sandur plain of the Pasterze glacier in Austria**

Kettle hole filled with water

J. Knight

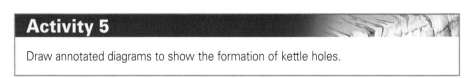

Activity 5

Draw annotated diagrams to show the formation of kettle holes.

Glacial overflow channels

Many glacial overflow channels or **spillways** formed in England at the end of the last glacial, approximately 13 000 years ago. Evidence of overflow channels can be seen in very steep-sided valleys, which are currently occupied by small underfit rivers (i.e. they are too small to have eroded the valley). The North York Moors is a classic area in which overflow channels formed. Perhaps the most famous are Newtondale and the Forge valley.

Towards the end of the last glacial, ice extended from continental Europe across the North Sea. As the climate warmed, the ice started to melt but the meltwater was unable to drain into the North Sea because the valley outlets were dammed by ice. Consequently, glacial lakes developed. Lake Pickering was the largest of these, occupying what is now called the Vale of Pickering.

To the north of Lake Pickering, another glacial lake, Lake Wheeldale, also filled with water, reaching a maximum height of 203 m above sea level. As more meltwater arrived, the lake level rose until it spilled over a low point of the valley side. Then huge volumes of meltwater drained to the south along Newtondale. As a result, the meltwater carved the Newtondale overflow channel. Today the dale is occupied by a very small river, Pickering Beck.

To the east of Newtondale, Forge valley was formed in a similar manner. Here Lake Hackness filled and overflowed to the south to form the steep-sided Forge valley (Figure 5.8).

Figure 5.8 **Location and formation of glacial spillways: Lake Pickering and the formation of Newtondale**

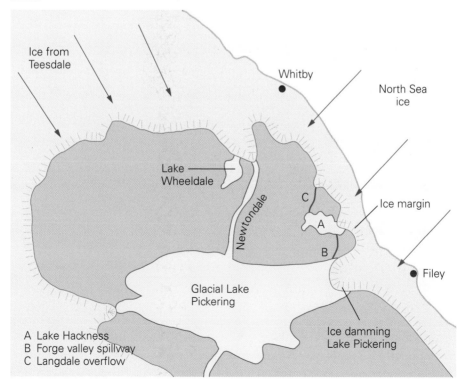

A Lake Hackness
B Forge valley spillway
C Langdale overflow

6 Periglacial processes and landforms

Periglaciation was the term introduced in 1909 by Walery von Loziński, a Polish scientist, to describe landforms and processes occurring around the margin of the Pleistocene ice sheets. Today the definition of periglaciation has been widened to include the areas found at the edge of contemporary glaciers or ice sheets (Figure 6.1). Periglacial areas can be found in:

- high-latitude areas of the northern and southern hemispheres that are not covered by ice
- mountainous areas, such as the Alps and the Rockies

| Figure 6.1 | **Distribution of contemporary permafrost areas in the northern hemisphere** |

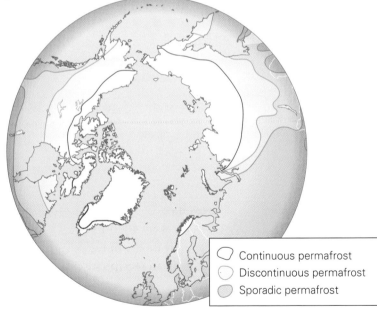

◯ Continuous permafrost
◯ Discontinuous permafrost
◯ Sporadic permafrost

Periglacial climates

In periglacial areas, part of the ground remains frozen for the entire year. This permanently frozen ground is known as **permafrost**. However, in response to higher temperatures in the summer, the surface layers of the soil thaw. A wide range of climatic zones experience periglacial conditions, from the extremely harsh continental interior of Siberia to the more maritime climates of Iceland and Spitsbergen.

The four types of region that experience periglacial conditions are shown in Table 6.1.

Table 6.1   Regions of the world that experience periglacial conditions

Region	Climate	Characteristics	Examples
Polar lowlands	Mean temperature of coldest month is <–3°C	Ice caps, bare rock surfaces, tundra vegetation	Northern Siberia, Arctic Canada
Subpolar lowlands	Mean temperature of coldest month <–3°C; mean temperature of warmest month >10°C	Taiga vegetation; the 10°C isotherm roughly coincides with the tree line in the northern hemisphere	Northern Europe, northern Canada
Mid-latitude lowlands	Mean temperature of coldest month is <–3°; mean temperature >10°C for at least 4 months of the year	Low-lying, undulating continental regions	Finland, central Siberia
Highlands	Climate influenced by altitude as well as latitude, so is highly variable; also affected by large diurnal temperature fluctuations; summer temperatures vary from –12°C to 10°C; precipitation is <30 cm per year	High mountainous areas above the tree line where vegetation is alpine or sparse	Kilimanjaro; the Alps; the Rockies

The climate of periglacial areas has:
- mean annual temperatures of between –1 and –3 °C
- temperatures below –10 °C for at least 6 months of the year
- low precipitation levels, with the mean precipitation being less than 1000 mm per year

The altitude at which permafrost is found increases closer to the equator. For example, Mauna Kea in Hawaii at 19 °N has sporadic occurrences of permafrost at 4170 m above sea level.

The low temperatures and low precipitation found in periglacial areas are not only related to their high latitudes and high altitudes but can also be explained by anticyclonic weather systems that cover periglacial areas for much of the year.

Activity 1

Explain why anticyclonic weather systems will lead to low temperatures and reduced precipitation in periglacial areas.

Activity 2

Table 6.2 shows climate data for Svalbard.

Table 6.2 Climate statistics for Svalbard

Month	Jan	Feb	Mar	April	May	June
Mean temp (°C)	−16.4	−16.8	−16.2	−13.1	−5.8	−0.1
Max temp (°C)	−12.6	−12.7	−12.4	−9.6	−3.4	2.1
Min temp (°C)	−20.3	−21.0	−20.0	−16.6	−8.2	−2.3
Precipitation (mm)	46	46	47	30	25	28

Month	July	Aug	Sept	Oct	Nov	Dec
Mean temp (°C)	3.5	2.7	−1.5	−7.0	−11.5	−15.2
Max temp (°C)	6.0	4.8	0.4	−4.7	−8.4	−11.6
Min temp (°C)	1.1	0.6	−3.4	−9.3	−14.6	−18.8
Precipitation (mm)	38	47	48	46	45	49

(a) Construct a climate graph to show the climate of Svalbard.
(b) Describe the climate of Svalbard.
(c) Calculate the range of mean annual temperatures and precipitation.
(d) Explain why precipitation is higher than you might expect.

Permafrost

Permafrost is permanently frozen ground where the surface layers melt during the summer. This melting layer or **active layer** can be up to 3 m thick (Figure 6.2). Its boundary with the permafrost is known as the **permafrost table**. Approximately 20% of the world's land surface is underlain by permafrost. There are three types:

- **Continuous permafrost** occurs in areas where the only unfrozen parts of the ground are below lakes, rivers or the sea.
- **Discontinuous permafrost** occurs where small scattered unfrozen areas appear.
- **Sporadic permafrost** occurs where small islands of permafrost exist in a generally unfrozen area.

Figure
6.2 **The energy budget required for formation of the active layer**

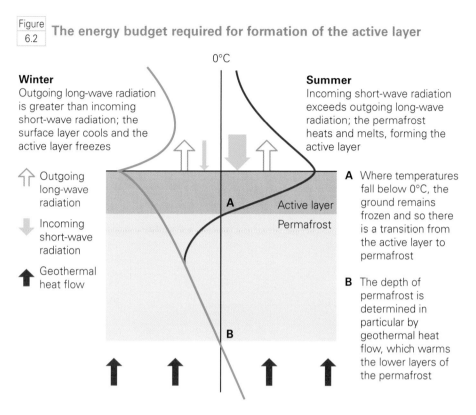

Winter
Outgoing long-wave radiation is greater than incoming short-wave radiation; the surface layer cools and the active layer freezes

⇧ Outgoing long-wave radiation

⬇ Incoming short-wave radiation

⬆ Geothermal heat flow

0°C

A

Active layer

Permafrost

B

Summer
Incoming short-wave radiation exceeds outgoing long-wave radiation; the permafrost heats and melts, forming the active layer

A Where temperatures fall below 0°C, the ground remains frozen and so there is a transition from the active layer to permafrost

B The depth of permafrost is determined in particular by geothermal heat flow, which warms the lower layers of the permafrost

The most extensive areas of continuous permafrost in the world are found in Canada, Alaska and Siberia (Table 6.3). Here the permafrost can extend to depths of 1500 m. Vast areas of the globe are covered by permafrost:

- The northern hemisphere has 7.64×10^6 km² of continuous permafrost.
- The northern hemisphere has 14.71×10^6 km² of discontinuous permafrost.
- Antarctica has 13.21×10^6 km² of continuous permafrost.
- Mountainous areas have 2.59×10^6 km² of discontinuous permafrost.

Table
6.3 Depth of permafrost in different parts of the world

Area	Depth of permafrost (m)
Markha River, USSR	1450–1500
Udokan, USSR	900
Yakutsk, USSR	198–250
Mackenzie Delta, Canada	18–366
Schefferville, Canada	>76
Churchill, Canada	30–61
Prudhoe Bay, Alaska	610
Fairbanks, Alaska	30–122

Activity 3

Study Figure 6.3 and explain how each of the factors shown can lead to changes in the rate of formation and depth of permafrost in periglacial regions.

Figure 6.3 **Factors that affect the depth and rate of permafrost formation**

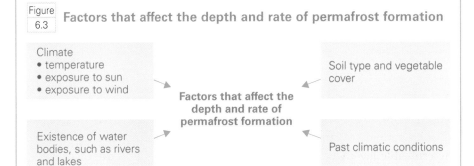

Climate
- temperature
- exposure to sun
- exposure to wind

Soil type and vegetable cover

Factors that affect the depth and rate of permafrost formation

Existence of water bodies, such as rivers and lakes

Past climatic conditions

Activity 4

Table 6.4 The latitude and depth of permafrost at various locations in Canada

Latitude	Depth of permafrost (m)
76.30	597
76.43	457
76.35	521
75.07	780
72.40	616
72.90	529
73.08	603
68.85	307
63.88	284
72.75	235
69.92	370
69.97	509
69.98	360
69.90	350
69.58	210
69.90	421
69.85	473
69.48	158
69.90	564
69.70	535

Activity 4 (continued)

(a) Use a suitable chart to present the data in Table 6.4.

(b) Calculate the Spearman rank correlation coefficient to measure the association between permafrost depth and latitude.

(c) What other factors might influence the depth of the permafrost?

Role of the active layer

The active layer is crucial to many processes and landforms found in periglacial areas. During the summer months, the ice in the surface layers melts but the subsoil (permafrost) remains frozen. Thus the permafrost at depth acts as an impermeable layer when the active layer has melted. The active layer refreezes with the onset of winter. Freezing occurs from the surface downwards and from the permafrost upwards. Occasionally some of the water remains unfrozen. This unfrozen soil is known as **talik** (Figure 6.4). Talik may form as a result of:

- the release of latent heat as the active layer freezes (closed talik)
- increased pressure as the freezing front encroaches on the talik (closed talik); the increased pressure means melting does not occur
- the layers immediately below a surface lake staying warm (open talik)

| Figure 6.4 | **Formation of talik due to increased pressure and the release of latent heat** |

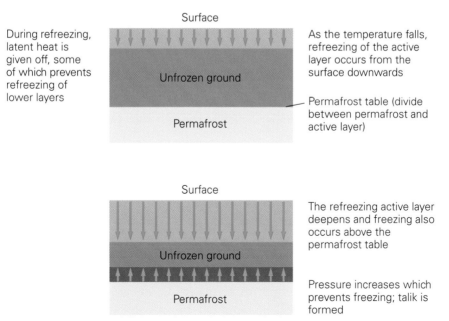

During refreezing, latent heat is given off, some of which prevents refreezing of lower layers

Surface

Unfrozen ground

Permafrost

As the temperature falls, refreezing of the active layer occurs from the surface downwards

Permafrost table (divide between permafrost and active layer)

Surface

Unfrozen ground

Permafrost

The refreezing active layer deepens and freezing also occurs above the permafrost table

Pressure increases which prevents freezing; talik is formed

Activity 5

Table 6.5
Latitude and thickness of active layers

Latitude	Active layer thickness (cm)
75.65	30
75.67	223
75.67	70
75.67	140
75.68	122
58.12	120
63.60	100
58.62	64
62.57	68
62.47	30
54.80	46
54.45	46
52.50	46
51.30	46
54.77	46
57.20	76
57.05	46
59.50	46
53.38	61
58.62	61

(a) Plot the data in Table 6.5 to show the relationship between latitude and the depth of the permafrost. Draw a best fit line.

(b) Calculate the Spearman rank correlation coefficient to show the strength of any relationship.

(c) Comment on the significance of the results.

(d) Use descriptive statistics to describe the thickness of the permafrost.

Processes in periglacial regions

Weathering processes

Periglacial areas are dominated by physical weathering. The most important of these processes is **freeze–thaw**. This occurs as an annual and a diurnal cycle. As temperatures fall, water trapped in the soil and rock joints freezes. Freezing causes a 9% volumetric increase as the water turns to ice. The repetition of this cycle and the pressure that is exerted on rocks and soil causes them to disintegrate. This is also known as **congelifraction**.

Chemical weathering processes are important too. **Hydrolysis** (the reaction of some minerals with water) occurs because of the abundance of water in summer months and the availability of acids released by vegetation. It is particularly frequent in granite areas, where the water mixes with feldspar. This produces potassium hydroxide, which is removed in solution. **Carbonation** also occurs because carbon dioxide is more soluble at lower temperatures. Carbonation is the reaction of rainwater, a weak carbonic acid, with the calcium carbonate of limestone. Calcium bicarbonate is produced, which is removed in solution.

The frost shattering of rock causes periglacial head deposits to be formed (weathered material that has moved due to solifluction). Figure 6.5 shows head deposits in Porlock Bay, Somerset, which experienced periglacial conditions during the penultimate glacial and as the ice sheet retreated.

Figure 6.5 **Head deposits, Porlock Bay, Somerset**

J. Knight

Mass movement processes

Gelifluction

Gelifluction is the main mass movement process in periglacial environments. It is the surface flow of **regolith** over a frozen subsurface. **Solifluction** is a similar process but it occurs only over a non-frozen surface. Gelifluction occurs because:

- soil particles are broken down during the winter months as the formation of ground ice forces them apart
- summer temperatures increase and the ground ice melts, increasing pore-water pressure
- the permafrost acts as the shear plane as it is a relatively impermeable surface
- any slight gradient in the permafrost will cause the saturated material to flow downhill

Given the relatively arid climate of periglacial areas, most of the soil moisture needed for gelifluction comes from melting ground ice. For example, some frozen silts in Alaska contain over 80% of ice by volume. Vegetation acts as an important restraining factor. Turf and moss can hold the semi-fluid layer in place until the load becomes too great and then gelifluction occurs, resulting in gelifluction lobes. The speed of gelifluction is variable (Table 6.6) and depends on the:

- amount of precipitation
- vegetation cover
- gradient of slope

Table 6.6 Rates of gelifluction

Place	Gradient (°)	Rate of movement (cm per year)
Spitsbergen, Norway	3	1.0
Spitsbergen, Norway	4	3.0
Spitsbergen, Norway	7–15	5.0–12.0
Kärkevagge, Sweden	15	4.0
Tarna area, Sweden	5	0.9–1.8
Norra Storfjell, Sweden	5	0.9–3.8
Okstindan, Norway	5–17	1.0–6.0
Ruby Range, Canada	14–18	0.6–3.5
Sachs Harbour, Banks Island, Canada	3	1.5–2.0

Frost heave

The expansion of water as it turns into ice occurs parallel to the greatest temperature gradient. As a result, expansion takes place normal to the ground surface. Thus as the water freezes, it expands and pushes stones to the surface. This process is called **frost heave**. During the melting season, as the ice thaws, finer material flows in to fill the space previously occupied by ice and so the stones remain supported. This is a slow process (Figure 6.6), with typical maximum speeds in the Canadian Arctic being 3.8 cm per month. Steeper slopes in Spitsbergen (at an angle of 7–15°) experience equally slow rates of 12 cm each year on average.

| Figure 6.6 | **Change in velocity of frost heave with depth** |

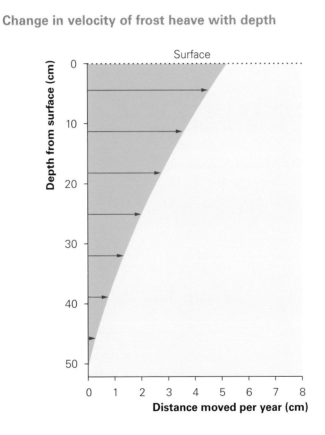

Frost creep

Frost creep also refers to the movement of individual stones (Figure 6.7). As with frost heave, when water turns to ice it moves normal to the ground surface. This time, however, as melting takes place, stones move downhill because of gravity as well as vertically to fill the vacuum created by melting ice.

Figure 6.7 **Frost creep velocity profile**

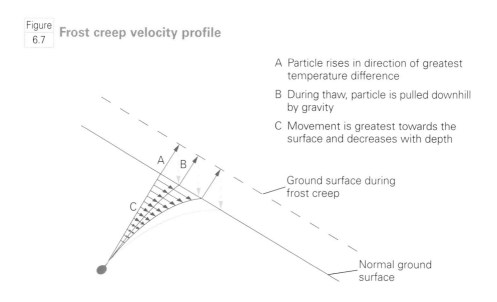

A Particle rises in direction of greatest temperature difference

B During thaw, particle is pulled downhill by gravity

C Movement is greatest towards the surface and decreases with depth

Ground surface during frost creep

Normal ground surface

Nivation

Nivation (altiplanation or cryoplanation) is a localised form of erosion that occurs because of the combined action of frost, gelifluction, frost creep and meltwater — for example, under snow patches. The snow provides meltwater but also insulates the ground beneath it. The thick snow cover reduces the amount of freeze–thaw. Northern Quebec, Canada, has some large nivation hollows, which may have been formed at a rate of $1500\,m^3$ per year.

Landforms in periglacial regions

Pingos

Pingos are isolated dome-shaped hills. Typically, their height ranges from less than 1 m to over 60 m and they can be up to 600 m in diameter. They develop in areas of continuous permafrost and are common in Alaska, Canada, Greenland and Siberia. Pingos are formed by two different processes:

- **Mackenzie-type pingos (closed system)** are found in the Mackenzie Delta, Canada and develop beneath lakes that are surrounded by permafrost. The soil beneath the central part of the lake is unfrozen as it is insulated by the water. Sediments are washed into the lake, which slowly infills, continuing the insulation. Over time, the water in the sediments freezes, but some trapped unfrozen material (talik) remains beneath this. As temperatures continue to decrease, the permafrost encroaches on this unfrozen material;

as it does, the pressure is increased due to water expanding on freezing. To relieve the pressure, the surface bulges upwards. Eventually all the water is converted to ice, forming a core of clear ice under the bulge (Figure 6.8). The initial stages can occur relatively quickly — up to 1.5 m per year in the first 2 years of formation.

Figure 6.8   **Formation of Mackenzie-type pingos (closed system)**

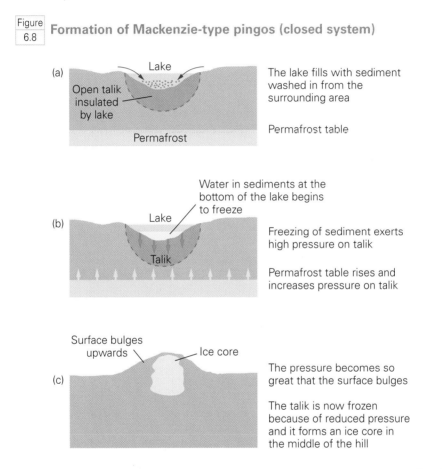

(a) The lake fills with sediment washed in from the surrounding area

Permafrost table

(b) Water in sediments at the bottom of the lake begins to freeze

Freezing of sediment exerts high pressure on talik

Permafrost table rises and increases pressure on talik

(c) The pressure becomes so great that the surface bulges

The talik is now frozen because of reduced pressure and it forms an ice core in the middle of the hill

- **East Greenland-type pingos (open system)** do not require a lake for their formation. As temperatures fall, there is progressive downward freezing of water-saturated sediments. Then, as subsurface pressure increases, it forces the ground to bulge upwards.

Activity 6

Draw a series of annotated diagrams to show the formation of East Greenland-type pingos (open system).

For both types of pingo, there may come a time when the intrusion of ice and the stretching of the overlying soil causes fracturing. This leads to the collapse of the pingo as the cracks allow warmer air to penetrate the ice core of the mound, and so it begins to melt. Thus, ruptured pingos (sometimes referred to as **ognips**) are characterised by a collapsed centre as the ice has disappeared. The longest known lifespan of a pingo is 1000 years.

Ice wedges

Ice wedges are widespread in periglacial environments. They develop in areas of continuous permafrost where the soils are poorly drained. During the winter, soil temperatures drop to below −15°C, causing the soil to contract and form cracks in the frozen ground. When temperatures rise during the spring, the soil expands and moisture collects in the cracks and freezes. Freezing causes the ice to expand and prevents the crack from closing. The average dimension of an ice wedge ranges from 15 m to 40 m (Figures 6.9 and 6.10).

| Figure 6.9 | Ice wedge at Criccieth, north Wales |

Coarse periglacial material

Fine-grained till

J. Knight

| Figure 6.10 | **The formation of ice wedges** |

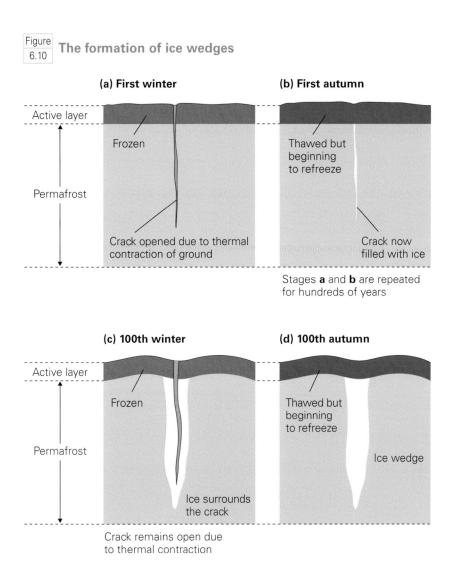

(a) First winter

Active layer

Permafrost

Frozen

Crack opened due to thermal contraction of ground

(b) First autumn

Thawed but beginning to refreeze

Crack now filled with ice

Stages **a** and **b** are repeated for hundreds of years

(c) 100th winter

Active layer

Permafrost

Frozen

Ice surrounds the crack

Crack remains open due to thermal contraction

(d) 100th autumn

Thawed but beginning to refreeze

Ice wedge

Patterned ground

Another distinctive feature of periglacial areas is **patterned ground**, where stones are sorted into regular patterns. Washburn (1979) classified patterned ground into five different types:

- **stripes** and **steps**, which occur in areas where there is a slope of between 5° and 30° (Figure 6.11)
- **polygons**, **circles** and **nets,** which are found on flat ground

Patterned ground does not form on slopes of over 30° as the processes of mass movement are too strong.

Figure 6.11 **Stone stripes on Skiddaw, Cumbria**

Parallel stone stripes

M. Raw

The larger, more-pronounced patterns are thought to be formed by cracking of the surface. The cracking can be caused by:

- contraction due to drying (desiccation)
- heaving and subsequent stretching of the soil (dilatation)
- thermal contraction due to the climatic regime of periglacial areas

Frost heave and dilatation are thought to be the primary causes of patterns. These processes segregate the larger stones, which move upwards and outwards from the finer matrix (Figure 6.12).

Over time, the patterns may join together and form polygons. However, the striping and steps found on slopes have a slightly different explanation. Their formation is associated with convection as melting takes place. Key to this process is the density of water, as in the following sequence:

- Water is at its densest at 4°C.
- Thawing takes place from the surface downwards during the summer months.
- A stage is reached when the relatively warm and dense water at 4°C sits above cooler but less dense water, which is closer to the freezing front of the permafrost.
- As there is a difference in temperature between different layers of water, convection begins. The less dense (but cooler) water rises to the surface, while the warmer but denser water begins to sink.
- This leads to localised melting of ice, so the surface of the frozen subsoil becomes uneven.

- The melting of ice beneath the surface results in pits in the overlying surface soil.
- Fine materials flow into the surface pits and coarser sediments remain on the raised areas between the pits.

Figure
6.12

The formation of patterned ground

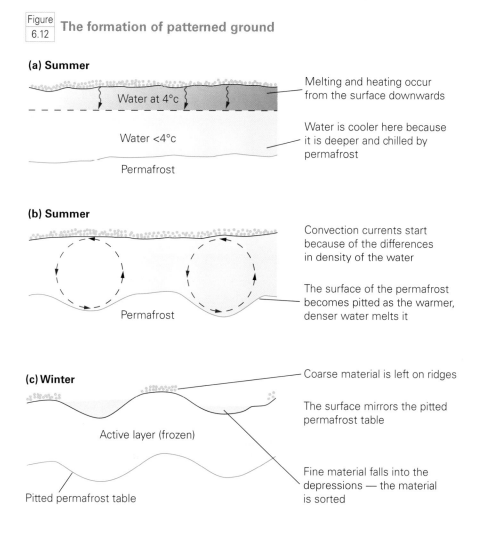

(a) Summer

Water at 4°c

Water <4°c

Permafrost

Melting and heating occur from the surface downwards

Water is cooler here because it is deeper and chilled by permafrost

(b) Summer

Permafrost

Convection currents start because of the differences in density of the water

The surface of the permafrost becomes pitted as the warmer, denser water melts it

(c) Winter

Active layer (frozen)

Pitted permafrost table

Coarse material is left on ridges

The surface mirrors the pitted permafrost table

Fine material falls into the depressions — the material is sorted

Thermokarst features

Thermokarst refers to a land surface that forms as ice-rich permafrost melts. Melting may be due to climatic warming or to an event such as a forest fire. **Alases** are large depressions that develop as the permafrost melts. Whatever the cause, the melting of the permafrost leads to large-scale subsidence and the formation of an alas or depression in the surface. Alases can be between 3 m and 40 m deep and between 100 m and 15 km in diameter.

Asymmetrical valleys

In many current or former periglacial areas, **asymmetrical valleys** develop, where one side of the valley is steeper than the other side (Figure 6.13). In the northern hemisphere, the north-facing slopes of the valley are usually steeper than the south-facing slopes. This is because of the differences in the amount of thawing each side of the valley experiences. The south-facing side has more thawing cycles and therefore more mass movement, in particular gelifluction.

| Figure 6.13 | The asymmetrical valley of the Dib in North Yorkshire. The valley is aligned from east to west and is composed of Carboniferous limestone. |

Steeper north-facing side of the valley

South-facing side of the valley is at a lower angle

J. Knight

Material moved down slope is often deposited as a lobe at the foot of the valley. There it may divert a river against the north-facing slope, leading to undercutting and hence retaining the slope's steepness. Vegetation is also an important factor. North-facing slopes generally have less vegetation cover and thus experience less mass movement due to the limited temperature change they experience.

Scree slopes

Extensive **scree slopes** are found in upland areas of the British Isles. Some are currently forming but most developed during periglacial conditions or towards the end of the last ice age. The principal process responsible for the formation of scree slopes is freeze–thaw weathering. This is most effective in well-jointed rock faces. Rock debris weathered by freeze–thaw accumulates at the base of the cliff. The most extensive screes in England are those found at Wastwater in the Lake District.

| Figure 6.14 | **Scree slopes caused by frost shattering at Cwm Idwal, north Wales** |

Well-jointed exposed rock free face

Scree accumulates at the base of the free face

J. Knight

Activity 7

Tables 6.7 and 6.8 contain scree measurement data from Upper Wharfedale. Scree was sampled randomly at 10 m intervals. Site 1 is at the base of the scree slope and site 6 is at the top of the slope.

Table 6.7 Slope angles at 10 m intervals

Site	1	2	3	4	5	6
Angle (°)	31	37	37	37	39	38

Table 6.8 Sample of particle sizes in cm (median axis)

Site 1	Site 2	Site 3	Site 4	Site 5	Site 6
15	9	15	14	9	8
18	12	19	10	10	9
12	10	9	10	8	9
10	5	9	12	12	8
14	10	13	14	11	8.5
7	10	14	12	12	13
8	12	13	15	6	7.5
6	7	19	7.5	9	11
10	8	24	7.5	9	11
15	9	8	9	10	8
7	8	10	13	5.5	5.5
9	11	15	16	13	12
12	8	6	6.5	11	10
7.5	10	13	16	7	11
9	10	10	15	11	2.5
13	11	13	16	9	7
10	9	6	14	9	7
10	10	11	15	8.5	11
10	12	11	14	14	10
12	11	6	11	11	4

(a) Use statistical methods to describe the size of scree particles for each site.

(b) Is there any obvious sorting pattern? Explain your results.

7 Human activities in glaciated and periglacial areas

Glaciated and previously glaciated areas have become important for certain types of economic activity, such as hydroelectric power (HEP) and tourism. This is because the physical environment of glaciated uplands is suited to such activities. U-shaped valleys and corries, high precipitation and relatively low temperatures provide ideal locations for HEP. Meanwhile, tourism has increased in areas such as the Alps and, more recently, Antarctica.

Case study: Hohe Tauern, Austria

Hohe Tauern National Park is one of four national parks in Austria; it is found in southern Austria to the south of Salzburg and covers the eastern section of the Alps (Figures 7.1 and 7.2). The park is the largest single protected area in the Alps. Grossglockner, the highest peak in Austria at 3798m, dominates the area. Because of the stunning mountain scenery found in Hohe Tauern, it is a popular tourist destination. All-year skiing is available where glaciers are found. Other recreational activities include walking, climbing, mountain biking and many extreme sports. The deep valleys on the edge of the park are also ideal for the generation of HEP.

| Figure 7.1 | Hohe Tauern National Park in the Austrian Tirol |

J. Knight

Figure 7.2 Typical alpine scenery in the Hohe Tauern National Park in the Austrian Alps

Energy production in Hohe Tauern

The glacial troughs and corries of the Hohe Tauern National Park are ideal locations for dam construction because these landforms are deep and relatively narrow. The area also has high precipitation (1500–2600 mm per year).

Wasserfallboden is an example of an HEP station that has been built at the edge of the Hohe Tauern National Park. It is located east of Kitzsteinhorn near the village of Kaprun. It has three reservoirs, two at 2000 m altitude (the Margaritze and Mooserboden storage lakes) and the Wasserfallboden (Figure 7.3), which is 360 m lower down. The water from the upper reservoirs falls through tunnels to the turbines at Wasserfallboden. The force of the water turns the turbines to generate electricity. During periods of low energy demand, the water is pumped back up to the higher lakes.

Figure 7.3 **Wasserfallboden reservoir in the Kaprun valley, Austria**

J. Woodhouse/Alamy

Tourism in Kitzsteinhorn

Glaciated and previously glaciated areas are often important tourist destinations. The Alps are popular during summer for walking and climbing and, more recently, for extreme sports. Most famously, however, the Alps are a hugely popular winter ski destination with more than 120 million visitors a year. The economic importance of tourism is shown by the following facts:

- Mountain tourism accounts for between 15% and 20% of all tourism revenue in the European Alps.
- Mountain tourism accounts for an estimated 7–10% of annual income from global tourism.
- The ski industry generates 4.5% of Austria's GNP.

In Austria, the Kitzsteinhorn ski area in the Hohe Tauern National Park (including the Pasterze glacier and Grossglockner Alpine Road) are both popular tourist destinations. The Kitzsteinhorn glacier in the Europa Sport Region of Austria is an all-year-round ski resort. Skiing started here in 1965 and today is of great economic importance to the two nearest settlements of Kaprun and Zell-am-Zee.

During its first 2 years as a ski resort, 200 000 people visited Kitzsteinhorn. The ski area provides 250 jobs, making it the largest employer in Kaprun. It now has 20 cable cars and lifts and 40 km of piste. The economic benefits of tourism in Kaprun are widely recognised. Tourism creates jobs, increases the disposable income of local residents and has a general multiplier effect.

However, there are environmental concerns over the sustainability of the ski industry. For example, the creation of ski pistes damages the mountain ecosystem, as large numbers of trees have to be cleared for the pistes, chairlifts and cable cars. Similarly, roads, tunnels and car parks need to be built, again requiring the clearance of vegetation. Aesthetically the pistes might look fine during the winter months when there is a good cover of snow but during the summer months they can be something of an eyesore.

One of the potential threats to the ski industry in Austria is global warming. The threat is particularly great here because 75% of the country's ski runs are situated below 1000 m.

Activity 1

Use webcams to compare the aesthetic qualities of a ski resort at different times of the year (winter, summer, autumn and spring). The website www.snoweye.com is a good start point. It has over 3000 webcams throughout the world and many webcams in the ski areas of the five main alpine countries.

Grossglockner Alpine Road

Grossglockner Alpine Road is a high-altitude road through glaciated landscapes in the Hohe Tauern National Park. The concept of such a mountain road was first considered in 1922 to link the village of Ferleiten with Heiligenblut. The road from Heiligenblut followed an earlier footpath and took 5 years to construct. It opened on 3 August 1935 and runs for 22 km at an altitude of more than 2000 m. Needless to say it is only open during the summer months when there is no snow!

Road engineers originally planned for approximately 120 000 visitors each year but 3 years after its opening some 375000 people were using the road each year, rising to 1.3 million in 1962. Today there is an average of 1 million visitors each year. Along the road there are a number of information centres where visitors can learn about the high altitude environment and the adaptation of the tundra vegetation to the climate of the area (Figure 7.4).

Figure
7.4 **The Grossglockner Alpine Road in Austria**

J. Knight

A side road leads to Franz Josefs Hohe, a large visitor centre perched above the Pasterze glacier and beneath the highest Austrian peak, the Grossglockner. The visitor centre offers guided tours of the area as well as an educational centre and information boards. The most popular activity at the visitor centre is a walk down steep steps to the Pasterze glacier, with a second walk through the proglacial area.

Activity 2

(a) Explain why high-altitude areas are so sensitive to visitors. Consider both the ecosystem and the climate.

(b) What might be the environmental impact on the Hohe Tauern National Park of more than 1 million visitors arriving in motor vehicles annually?

Given the number of visitors to the Hohe Tauern National Park each year and the fact that it is an all-year destination, there are, inevitably, pressures on the environment, which means that careful management of the National Park is needed. Alpine ecosystems are fragile, so scientific research stations have been set up within the park, particularly to look at the tundra ecosystem.

Sustainable use of Hohe Tauern

The number of visitors is of great economic value to the population of the Hohe Tauern National Park and to Austria. However, tourism places immense pressure on the National Park, meaning that potentially its use could be unsustainable. The aims of the National Park are to:
- encourage protection of the environment
- facilitate scientific research
- encourage tourism
- educate people about their environment

It is difficult to achieve a balance between promoting use of the National Park and conserving the environment; hence tourism has several noticeable impacts within the National Park, including:
- erosion of footpaths on popular walking routes, particularly in more accessible areas
- erosion of pistes during the ski season
- use of snow cannons to generate snow, lengthening the ski season but delaying the spring melt and thus affecting the flora
- deforestation to facilitate the construction of ski pistes
- pollution
- increase in building nearby

So, what is being done to ensure that this beautiful, varied and pristine landscape is not damaged irreversibly? There are several strategies that have been put into place in an attempt to ensure the long-term future of the National Park:
- Footpaths are maintained and well signposted, with hiking maps available freely to hikers (Figure 7.5).
- The National Park has been divided into a core area, which is relatively pristine, and a peripheral area, which is permanently populated.
- Sustainability is being encouraged in settlements.
- The National Park is a member of the Network of Alpine Protected Areas, a scientific network established between National Parks to protect the natural environment.

Figure 7.5　**Management of tourism in the Austrian Alps**

Signposts show hikers the routes, their grade and the distance to the end of the route

Schleinitz

Zettersfeld Zentrüm

Sessellift · Steinermandl

J. Knight

An example of a sustainable tourist resort is Neukirchen on the edge of Hohe Tauern. This small village is popular with both summer and winter tourists. It has a permanent population of 2600 and it can accommodate 3300 visitors during the tourist season. The ski area at Neukirchen is relatively small and therefore is placed under great pressure during the peak season. So what is done to ensure that Neukirchen and the surrounding area adjacent to the National Park are used sustainably?

To minimise the impact of skiing, the more vulnerable sections of the slopes are covered with straw. The snow then falls onto this and the piste is compacted. This protects the slope and so reduces the impact of skiing and erosion. During the summer, these slopes are grazed by cows, which maintain the biodiversity but also replenish nutrients in the soil.

One of the ski lifts in the Wildkogel ski area is driven by solar energy, a renewable energy source, while the town itself is self-sufficient for energy with its own supply of HEP, another renewable energy source. The hotels are relatively small, while the restaurants use local produce, also in an attempt to ensure that the tourism is sustainable.

Tourism in Antarctica

Antarctica is often described as the last great wilderness because of its remoteness and extreme environment. How true is this? It has been described as '… a natural reserve, devoted to peace and science' (Antarctic Treaty parties). The Antarctic Treaty, signed in 1959 by 12 nations, outlines appropriate uses of the continent:

- Military activities are not allowed.
- Scientific research is permitted.
- Nuclear testing and waste disposal are not allowed.
- Work can be inspected at any time.
- Expeditions and travellers to the continent should give notice of their intention to visit.

Today there are over 60 research stations in Antarctica belonging to 20 nations. The continent has a summer population of about 2400, which reduces by half in the winter months. Seven countries have made territorial claims to the continent, including the UK. However, thanks to the 1959 Antarctic Treaty, the continent is not owned by any one country. What is unique is that 45 countries, which between them have 80% of the world's population, have signed the Antarctic Treaty and are committed to the preservation of Antarctica.

Figure 7.6 **Map of Antarctica**

History of tourism

International tourism has increased massively since the first package holiday was sold in 1956. Antarctica is the latest destination to become available for tourism. People arrive either by aircraft or by boat, with most visitors wanting to see the Antarctic wildlife. There is currently no accommodation for tourists on Antarctica and the tourist season is limited to the summer months when there is up to 23 hours of daylight and temperatures hover between −4 °C and +4 °C. Most passengers are Australian, American or British. The television programme *Life in the Freezer*, broadcast in 1993, led to a surge of British tourists visiting the continent.

Tourists first visited Antarctica in 1958. Their experience was limited to a flight over the Antarctic Peninsula. It was not until the 1960s that commercial flights landed at McMurdo Sound and at the South Pole. In the 1950s, the first tourist ship arrived in Antarctica with 100 passengers onboard an Argentinian vessel. Larger ships carrying 500 people are now standard. Helicopter trips from the ships land tourists on the Antarctic Peninsula. In 2002/03, 74% of all visitors arrived onboard ships that landed on the Antarctic coast.

The number of visitors, although small, is increasing rapidly (Figure 7.7). Annual tourism numbers grew from 4800 visitors in 1990 to 16 000 in 2003. Anticipated rates of increase suggest that approximately 26 000 tourists could visit Antarctica during the 2006/07 season. Of all the travellers to Antarctica, 90% will set sail from Ushuaia, the southernmost city of South America, at the tip of Argentina. From there, the ships sail through Drake's Passage and then across the sixtieth parallel (the area south of this line of latitude is the official definition of Antarctica).

 Figure 7.7 **Number of visitors to the Ross Sea and other parts of Antarctica**

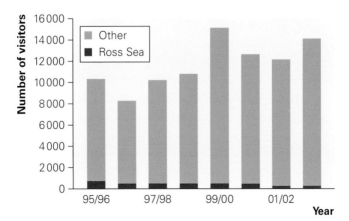

Activity 3

Figure 7.8 The types of visit made to the Ross Sea and other parts of Antarctica in 2002/03

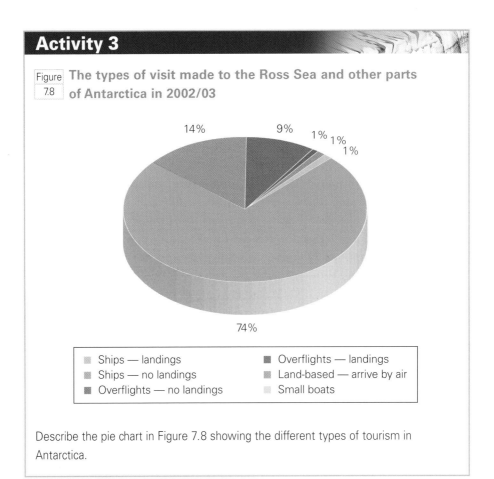

14%   9%   1% 1%
                 1%

74%

- Ships — landings
- Ships — no landings
- Overflights — no landings
- Overflights — landings
- Land-based — arrive by air
- Small boats

Describe the pie chart in Figure 7.8 showing the different types of tourism in Antarctica.

Impact of tourism

There is growing concern that the scale of tourism in Antarctica is increasing with the emergence of large, general-purpose ships operated by global corporations. There is currently only one icebreaker tourist ship that is equipped to withstand the harsh weather conditions of the Antarctic. This is a 12 000-tonne Russian icebreaker, the *Kapitan Khlebnikov*.

Similarly, there has been an increase in the number of 'adventure' activities available in Antarctica. Tourists are visiting evermore remote sites, such as the Pole of Inaccessibility and the Dry Valleys discovered by Scott in 1903. There is currently only one travel company that offers accommodation in Antarctica: Adventure Network International. Not surprisingly, given the pristine nature of the continent and the fragility of its ecosystems, there is some concern over the footprint of tourism and its sustainability.

Management of tourism

What is being done to minimise the effect of tourism on Antarctica? Each travel company involved in Antarctica has to carry out an environmental impact study to ensure that its effect is minimal. This is submitted to the International Association of Antarctica Tour Operators (IAATO), an organisation formed in 1991 by seven tour companies. IAATO restricts access to the most sensitive sites and limits the number of tourists that go ashore.

However, travel companies do not have to belong to IAATO in order to visit Antarctica, so its initiatives are somewhat limited. Tourists landing on Antarctica have to disinfect their footwear as they leave their boat and as they board. There are no toilet facilities at the tourist locations and tourists are not supposed to walk within 5 m of any wildlife.

Many organisations think that these restrictions are inadequate and want stricter controls to ensure that Antarctica is not overrun with tourists. The Antarctic and Southern Ocean Coalition (ASOC) has made the following suggestions:

- A limit should be set on the overall level of tourist activity in Antarctica over the next 20 years. This would be a limit not only on the number of tourists but also on how people arrive, how they are transported through Antarctica, and the sites they visit.
- Land-based accommodation and infrastructure should not be developed.
- Air transport of tourists to land on Antarctica should not be allowed.
- Large, general ocean-going vessels should be banned from the Southern Ocean and only ice-breaking vessels should be allowed.
- Baseline information about the more remote sites should be collected by the tour operators.
- Tour operators should not be allowed to proceed with visits authorised by national authorising bodies if their environmental risk assessment indicates that a visit would be detrimental to the environment.

Activity 4

(a) Suggest reasons for the strategies put forward by ASOC for the control of tourism in Antarctica.

(b) What other strategies could be imposed to minimise the environmental impact of tourism.

(c) Rank the strategies in order of effectiveness. Give reasons for your decisions.

Future of glaciated areas

There is much evidence from glacial studies to show that many glaciers are currently retreating and that the main cause is a rise in atmospheric temperatures due to human activity. In the first half of the twentieth century in the northern hemisphere, atmospheric temperatures rose by $0.3\,°C$, due largely to the increase in levels of carbon dioxide. This warming has pushed many glaciers into negative mass balance — hence they are retreating and the volume of ice they contain is reducing. For example:

- The eastern slopes of the Rocky Mountains in the USA have 1300 glaciers that have lost between 25% and 75% of their mass since 1850.
- In 1949 in Tajikistan in central Asia, glaciers covered $18\,000\,km^2$ compared with $11\,863\,km^2$ in 2000. This is a 35% decrease in the area covered by ice.
- Some Himalayan glaciers are retreating at a rate of 10–15 m per year.
- 95% of glaciers in the Himalayas are retreating.
- The Khumbu glacier at the base of Everest has retreated more than 5 km since 1953.
- The area of Peru covered by glaciers has shrunk by 25% over the past 30 years.

There is much literature to show that the majority of glaciers throughout the world are experiencing retreat (Figure 7.9). What are the reasons for this?

- During the twentieth century, global temperatures rose by $0.6\,°C$.
- The 1990s was the warmest decade since record-keeping began in the 1800s, with the hottest years recorded being 1997, 1998, 2001, 2002 and 2003.
- 2006 was the sixth warmest year on record, exceeded by 1998, 2005, 2003, 2002 and 2004.
- In the northern hemisphere, summer is longer than it was 150 years ago, spring is arriving 9 days earlier and the winter freeze starts 10 days later.
- Less ice on the Earth's surface lowers the albedo, so there is reduced reflection and more warming of the Earth's surface and atmosphere, and hence more melting.

There are, however, a few regions where glaciers are expanding — for example, the maritime glaciers of Scandinavia. Some regions in the Himalayas are also experiencing glacier growth.

What are the consequences of this warming for the Earth's human population? With the melting of icecaps and glaciers, global sea-level is set to rise, with a typical eustatic rise of up to 88 cm expected, although melting of the entire Greenland ice sheet alone could cause a 7 m rise in sea level and complete melting of Antarctica would cause a 60 m rise.

Figure 7.9	**Retreat of the Muir glacier, Alaska, between 1976 and 2003**

B. Molnia (1976) *Muir Glacier*: from the *Online glacier photograph database*, Boulder, Colorado USA: National Snow and Ice Data Center/World Data Center for Glaciology: digital media

(a) Muir glacier, Alaska, in September 1976

B. Molnia (2003) *Muir Glacier*: from the *Online glacier photograph database*, Boulder, Colorado USA: National Snow and Ice Data Center/World Data Center for Glaciology: digital media

(b) Muir glacier, Alaska, in August 2003, taken from the same point

Threats to water supply

Another serious consequence for millions of people is the threat to their water supply. In mountainous areas, such as the Andes and Himalayas, glacial meltwater feeds rivers that supply small rural settlements. Furthermore, major rivers in Asia, such as the Mekong, Yangtze, Brahmaputra, Ganges and Yellow River, are fed by glacial meltwater. Changes in the hydrological cycle of these drainage basins have major implications for the world's two most populated countries: China and India. Initially there may be an increase in the discharge of the rivers, but this would be followed by a decline.

| Figure 7.10 | **Location of places and rivers that are important for water supply in Asia** |

Western China is a desert area with 300 million farmers dependent on the water supplied by glacial meltwater from the Qinghai–Tibetan plateau. Of 5000 glaciers monitored on the plateau, 82% had retreated by 4.5% over the past 50 years. Further harrowing evidence of glacial retreat can be seen in the Qilian Mountains, where 95% of the 170 glaciers are thinning by up to 4.9 m every year.

In western China there is less water flowing in rivers than in previous years because higher temperatures have led to increased evaporation of the meltwater from glaciers. Such water shortages could affect 42% of China's population — 538 million people in total. The reduction of glacial meltwater flowing into the Ganges is anticipated to result in at least 500 million people facing water shortages and 37% of India's irrigated land being affected.

Peru also faces similar threats and challenges. Here the coastal strip of Peru is arid, with little or no rainfall every year. The capital city of Lima is part of this strip, with a population of 8 million people. The rivers that supply water to western Peru and allow half of Peru's agricultural productivity to take place may well dry up in 20 years. With glacial retreat, summer melting will stop and the rivers will not flow for half of the year. In anticipation of this problem, the Peruvian authorities are investing in dams to store water for their population. However, what will happen if the glacier meltwater is no longer produced at all?

Activity 5

Use the internet to research the effects that melting glaciers might have on human activities, such as the skiing industry, HEP production and agriculture. Look for facts and figures to support your findings.

Human impact on periglacial environments

Periglacial areas typically have low population densities with many indigenous communities. However, many periglacial areas are being developed because of the richness of their natural resources. Both the environmental impact caused by the unsustainable exploitation of resources and the global threat of climate change have serious implications for periglacial areas.

Oil extraction in Alaska

Human activity has increased rapidly in some periglacial environments over the past 30 or 40 years. For example, Prudhoe Bay in northern Alaska has been transformed from a small Inuit settlement to a wealthy town where the economy is based on oil extraction. This development has not been without environmental cost. The Trans-Alaska pipeline, which transports oil from Prudhoe Bay to the Pacific Ocean at Valdez, has been designed to minimise its impact on the permafrost. It is built above the ground so that the oil, which is heated to keep it flowing, does not melt the permafrost (Figures 7.11 and 7.12). The oil pipe is also insulated so that, again, the heat from the oil does not disturb the natural environment.

Impact

Figure
7.11

The Trans-Alaska oil pipeline from Prudhoe Bay

Built in sections that are able to move; the sections are insulated to prevent the oil from freezing

Raised above ground level to avoid melting the permafrost

T. Eckmann

Figure
7.12

Location of the Trans-Alaska oil pipeline

Arctic Ocean

Prudhoe Bay oilfield

The Arctic National Wildlife Refuge

Trans-Alaska oil pipeline

ALASKA

CANADA

Anchorage

Valdez

Juneau

Pacific Ocean

0   km   500

Oil extraction is also taking place in the periglacial environment of Norman Wells, Canada. Here the pipeline is buried within the permafrost. The pipeline, completed in 1985, is 869 km in length; it transports oil from Norman Wells to Alberta. Unlike the oil transported across Alaska, this oil is chilled to −1 °C as it enters the pipeline. Permafrost slopes along the pipeline are insulated with woodchips to prevent an increase in slope failure as the permafrost thaws. Despite few obvious impacts on the permafrost, a forest fire in 1994 along the pipeline route led to active-layer detachment landslides, which caused severe damage to the pipeline.

Activity 6

The US government is dependent on oil reserves from Alaska to reduce its energy deficit. It wants to increase the total amount of natural resources extracted from Alaska. This is a controversial issue, based on economics versus the environment. Research the contrasting views using the internet. The following websites are useful starting points:

www.news.independent.co.uk/environment/article351121.ece

www.fema.gov/plan/prevent/earthquake/sty_oil.shtm

pubs.usgs.gov/fs/2003/fs014-03/pipeline.html

www.anwr.org

www.labor.state.ak.us/trends/nov06.pdf (pages 1–7 of this lengthy report)

www.guardian.co.uk/flash/0,,534962,00.html

Tourism

Tourism imposes further pressures on periglacial environments. These fragile areas have a very low carrying capacity: any increase in the number of visitors inevitably leads to some environmental damage. In northern Quebec, an Inuit town called Kuujjuaq serves as a gateway for hunting and fishing holidays. While visitor numbers are not high, the impact of tourists walking through the tundra has caused irreversible damage. Tuktoyaktuk in Northwest Territories, Canada, is the Inuit town with the greatest concentration of Mackenzie pingos (1400 in total); this area has now been given landmark status by the Canadian government to help protect the landforms and area from tourists.

Climate change and periglacial regions

The periglacial environment is a fragile system. The presence of permafrost is dependent on a finely tuned energy budget. This means that periglacial regions are vulnerable to human activity. Any changes, including increases in atmospheric temperature, clearing of vegetation, removal of the insulating organic layer and forest fires, alter the thermal regime of the ground and melt the permafrost.

The Mackenzie valley and delta are being monitored closely by the Geological Survey of Canada. Significant atmospheric warming has occurred there over the past 10 years, with a temperature increase of $1.7\,^\circ$C. General circulation models predict that a doubling of atmospheric carbon dioxide produced by human activity will cause temperatures in the Arctic, including the Mackenzie valley and delta, to increase by $3–5\,^\circ$C by 2100. With such temperature increases, areas of discontinuous and sporadic permafrost are likely to disappear. Deep continuous permafrost is only likely to be affected over a timescale of hundreds to thousands of years.

Areas with a high ice content in the permafrost are likely to suffer most because melting of the permafrost will lead to subsidence and permafrost degradation. This, in turn, will have a significant effect on infrastructure and economic activities. In July 2002, inhabitants of the Inuit village of Shishmaref abandoned their homes and moved 8 km inland because of rapid coastal erosion, caused in part by melting of the permafrost. Alaska is now $2\,^\circ$C warmer in summer and $4\,^\circ$C warmer in winter than it was 30 years ago.

Vast areas of permafrost in Alaska and western Siberia are showing evidence of greater surface melting in the summer months than has been previously observed. In Siberia, an area of 1 million km^2 (an area the size of France and Germany combined) has started to melt. This is the first significant melting known for the past 11 000 years. The melting has only really occurred in the last 3–4 years and is due to the $3\,^\circ$C warming recorded in western Siberia since the mid 1960s.

Melting leads to melting

The partially melted permafrost compounds the melting process. The bare ground alters the radiation budget as it has a lower albedo than the ice-filled permafrost. This increases the absorption of incoming short-wave radiation and raises surface temperatures. Hence melting of ice in the permafrost increases. The increase in air temperature and darker surfaces mean the formation of more ice wedges. During the summer, water flows into the depressions left by the ice wedge top and lakes develop. Over time, the lakes join together. Eventually, when the permafrost beneath the lake melts, the lakes will drain away.

Do periglacial areas increase global warming?

While a change in the nature of many periglacial areas can be observed, a real threat from the destruction of the permafrost is its little talked-about release of massive amounts of greenhouse gases.

Tundra areas are vast frozen peat bogs and the permafrost is a huge store of greenhouse gases. Melting will release vast amounts of methane (a potent greenhouse gas) and possibly carbon dioxide into the atmosphere. One estimate suggests that 70 billion tonnes of methane (25% of the world's terrestrially stored methane) is currently locked up in permafrost. It could potentially double the atmospheric level of methane, leading to a 10–20% increase in global warming. Where melting is underway, methane hotspots have been found in western Siberia, where methane bubbles to the surface and prevents refreezing of the active layer.

Activity 7

Using the BBC website, Telegraph online and other newspapers, research the threat of global warming to communities such as the Inuit. What changes have already taken place in their natural environment and how have they adapted their lifestyle to accommodate global warming? The following websites are a good place to start:

www.guardian.co.uk/international/story/0,3604,1104241,00.html

news.bbc.co.uk/2/hi/americas/3308827.stm